OLIVE

Improvement, Production and Processing

The Authors

Mr. Shiv Lal is presently Scientist (SS) at ICAR-Central Institute of Temperate Horticulture, Srinagar, J&K. He did his B.Sc. (Agriculture) from Govind Ballabh Pant University of Agriculture and Technology, Pantnagar, M.Sc. (Fruits Science) from Indian Agriculture Research Institute, New Delhi and currently pursuing Ph.D. (Fruit Science) from Indian Agriculture Research Institute, New Delhi. From the last 7 years he is actively engaged in olive breeding, production and processing research. He has established olive research block, evaluated large numbers of germplasm introduced from Egypt and UC, Davis, California and identified suitable genotypes for temperate region. He has also worked on olive seed germination, propagation, pollination management, maturity indices and oil quality and formulated several international cooperation proposals on olive germplasm exchange. He is recipient of ICAR NTS, JRF and SRF and joined ICAR services in 2008. He has published more than 40 research papers and contributed 5 bulletins, 5 book chapters, besides 5 presentations in national and international conferences. He has been awarded best research paper from The Horticultural Society of India.

Dr. M.K. Verma is presently the Principal Scientist at the Division of Fruits and Horticultural Technology, ICAR-Indian Agricultural Research Institute (IARI), New Delhi and involved in research & teaching. He received Ph.D. degree in Horticulture (Fruit Science) from IARI, New Delhi and joined Agricultural Research Services in 1998. Previously he has served at ICAR-Central Institute of Temperate Horticulture, Srinagar/Mukteshwarfor 12 years. He has significantly contributed in the field of temperate fruits in general and olive in particular. Worked for a short period on olive in Egypt under Indo-Egypt work plan and introduced the olive cultivar from Horticultural Research Institute, Cairo. He has developed 23- varieties of walnut, apricot, sweet cherry, apple, grapes, sweet orange and acid lime. He has published 39 research papers, 2- books, 45 book chapters, 36 conference papers, and 23 bulletins/folders.

Dr. Nazeer Ahmed is currently the Vice Chancellor, SKUAST (K), Srinagar and former Director, ICAR-Central Institute of Temperate Horticulture (CITH), Srinagar, J&K. He had his graduation in Horticulture from UAS Bangalore and Master's and Doctoral degree from PAU, Ludhiana. During his 30 years of service he guided more than hundred Master's and Ph.D's and published more than 300 research papers, extension bulletins and review papers. Dr Ahmed developed, released and popularized 33 varieties and hybrids and recommended 64 production recommendations for temperate fruits and vegetables. He received Rajbhasha Gaurav award, Dr. R.S. Paroda Award; Dr. M.H.Marigowda National Endowment Award; Dr. Kirti Singh Gold Medal; Vijay Shree Award; Glory of India Gold Medal and Chaudhary Devilal Outstanding award for best AICRP (VC & Potato).

Dr. Desh Beer Singh is presently Director/Acting at ICAR-Central Institute of Temperate Horticulture, Srinagar, J&K. He did his M. Sc (Horticulture) from Punjab Agricultural University, Ludhiana, Punjab and Ph.D. (Horticulture) from Bhim Rao Ambedkar University, Agra. Further he has obtained advanced training from USDA, USA in the field of conservation of plant genetic resources and from RAS, Moscow (Russia) in the field of postharvest management of horticultural produce and also visited Central Asian countries (Kazakhstan, Kyrgyzstan, Tajikistan, and Uzbekistan) for discussion on potential for collaboration research between Central Asian scientific institutions and ICAR on diversity of temperate fruit trees. He has published more than 110 research papers and contributed 14 bulletins, 15 book chapters, 06 books besides 57 presentations in national and international conferences.

OLIVE
Improvement, Production and Processing

– *Authors* –

Mr. Shiv Lal
Scientist (SS) at ICAR-Central Institute of Temperate Horticulture, Srinagar, J&K.

Dr. M.K. Verma
Principal Scientist (Horticulture-Fruit Science), Division of Fruits and Horticultural Technology, Indian Agricultural Research Institute, New Delhi

Dr. Nazeer Ahmed
Vice Chancellor, SKUAST (K), Srinagar & Former Director, ICAR-Central Institute of Temperate Horticulture (CITH), Srinagar, J&K

Dr. Desh Beer Singh
Director/Acting at ICAR-Central Institute of Temperate Horticulture, Srinagar, J&K.

2016

Daya Publishing House®
A Division of
Astral International Pvt. Ltd.
New Delhi – 110 002

Publisher's Note:

Every possible effort has been made to ensure that the information contained in this book is accurate at the time of going to press, and the publisher and author cannot accept responsibility for any errors or omissions, however caused. No responsibility for loss or damage occasioned to any person acting, or refraining from action, as a result of the material in this publication can be accepted by the editor, the publisher or the author. The Publisher is not associated with any product or vendor mentioned in the book. The contents of this work are intended to further general scientific research, understanding and discussion only. Readers should consult with a specialist where appropriate.

Every effort has been made to trace the owners of copyright material used in this book, if any. The author and the publisher will be grateful for any omission brought to their notice for acknowledgement in the future editions of the book.

Cataloging in Publication Data--DK
Courtesy: D.K. Agencies (P) Ltd. <docinfo@dkagencies.com>

Shiv Lal, author.
Olives : improvement, production and processing / authors, Mr. Shiv Lal, Dr. M.K. Verma, Dr. Nazeer Ahmed, Dr. Desh Beer Singh.
pages cm
Includes bibliographical references and index.

ISBN 978-93-86071-04-0 (International Edition)

1. Olive. I. Verma, M. K. (Horticulturist), author. II. Ahmed, Nazeer (Horticulturist), author. III. Desh Beer Singh, author. IV. Title.

SB367.S55 2016 DDC 634.63 23

Published by : **Daya Publishing House®**
A Division of
Astral International Pvt. Ltd.
– ISO 9001:2008 Certified Company –
4760-61/23, Ansari Road, Darya Ganj
New Delhi-110 002
Ph. 011-43549197, 23278134
E-mail: info@astralint.com
Website: www.astralint.com

भारत सरकार
कृषि एवं किसान कल्याण मंत्रालय
कृषि, सहकारिता एवं किसान कल्याण विभाग
कृषि भवन, नई दिल्ली- 110001

Government of India
Ministry of Agriculture and Farmers Welfare
Department of Agriculture, Cooperation and Farmers Welfare
Krishi Bhawan, New Delhi- 110001

Foreword

Olive (*Olea europe* L.) a tree borne oil seed crop, is primarily grown for its oval shape fruit which is widely used for extracting edible oil and also eaten raw in pickles, salads, soups, fast foods etc. It is valued for the presence of polyunsaturated fatty acids and is absolutely free from cholesterol. Owing to such properties, the consumption as well as trade has increased manifold in India. India entirely depends on the imports and spending approximately Rs. 350 lakh on annually basis (APEDA, 2015). It is commercially grown in Mediterranean countries. Considering the importance of the crop, an Indo-Italian project for promotion of olive was implemented in 1984 at two phases from 1984-87 and 1990-93 at all the three Northern hilly states of India *viz.*, J&K, H.P. and former U.P. hills (Now Uttarakhand) . Recently a pilot project on 250 hectares of land with 1.25 lakh olive trees of three varieties in dry desert of Rajasthan has been planted. The initial results under Bikaner (Rajasthan) conditions showed some promise, but the overall results are yet to come.

Wide array of technologies have been evolved around the world on olive production related aspects and there is lack of literature in book form at one place. Therefore, authors has made an effort for compilation of available knowledge and abridged in a book form. The publication contains over 25 chapters on modem technological inventions related to crop improvement, fruit production, processing, value addition and management of biotic and abiotic factors. I expect that this book will evoke enthusiasm for olive, which I know is infectious. I congratulate authors Dr. Shiv Lal, Dr. M.K. Verma, Prof. Nazeer Ahmed and Dr. D.B. Singh for their hard work in bringing out this publication. I am sure thal this publication would be immense helpful to researchers, academicians, policy makers, students, farming community and other stakeholders.

(S.K. Malhotra)
Agriculture & Horticulture Commissioner

Preface

Olive (*Olea europaea*) is undoubtedly one of the world's oldest cultivated crops of Mediterranean region but grows well even under mild temperate conditions if chilling requirements are met. Therefore, it has spread in many new areas around the world. The olive is not only a significant food source but also contributes to human health. Over 90 per cent of its fruits are utilized for extraction of oil and on a limited scale they are used for table purpose and in making pickles and in salad. Its oil is free from cholesterol having poly unsaturated fatty acids, rich in antioxidants and is consumed in large quantity mainly for cooking. It is considered useful in regulating high blood pressure and is safe for patients having coronary heart disease. Olive oil good for human skin and its massaging gives relief from muscular pains and provides warmth during cold weather. Due to much awareness about the health benefits associated with olive beside crop diversification, planners, researchers and other stake holders of India emphases the need of olive production in India. The initial experiments were conducted in North Western Himalayan states (H.P. and J&K) and in recently, Rajasthan also entered in olive production. However, the production of olive in India did not achieve the economic thresholds, mainly associated with large number of factors. Therefore, this book has been written.

This book of "olive" is the result of many years' endeavours in collecting valuable information from existing literature concerning the olive tree and its culture; from its history, through traditional practices to modern techniques and horticultural procedures. Furthermore, this book covers the basic physiological, biochemical, breeding and horticultural principles of olive culture.

Our objective is to provide the knowledge approprate for reserchers, academicians, students, planners, and growers, both experienced and inexperienced horticulturists and, in general, for anyone wishing to obtain knowledge and experience of olive culture in order to increase productivity and improve product quality. The content is written in a style accessible to anyone and is also appropraite as a textbook at various levels of education. To enhance understanding the text includes illustrations, figures and tables featuring experimental data, and it is a recent addition to the literature on 'olives'.

Mr. Shiv Lal

Dr. M.K. Verma

Prof. Nazeer Ahmed

Dr. D.B. Singh

Contents

1
History

The olive tree believes to originated to Asia Minor and spread from Iran, Syria and Palestine to the rest of the Mediterranean basin 6,000 years ago. It is among the oldest known cultivated trees in the world being grown before the written language was invented. It was being grown on Crete by 3,000 BC and may have been the source of the wealth of the Minoan kingdom. The Phoenicians spread the olive to the Mediterranean shores of Africa and Southern Europe. Olives have been found in Egyptian tombs from 2,000 years BC. The olive culture was spread to the early Greeks then Romans. As the Romans extended their domain they brought the olive with them. 1,400 years ago the Prophet of Islam, Muhammad, advised his followers to apply olive oil to their bodies, and himself used oil on his head. The use of oil is found in many religions and cultures. It has been used during special ceremonies as well as a general health measure. During baptism in the Christian church, holy oil, which is often olive oil, may be used for anointment. At the Christmas mass, olive oil blessed by the bishop, "chrism", is used in the ceremony. Like the grape, the Christian missionaries brought the olive tree with them to California for food but also for ceremonial use. Olive oil was used to anoint the early kings of the Greeks and Jews. The Greeks anointed winning athletes. Olive oil has also been used to anoint the dead in many cultures. The olive trees on the Mount of Olives in Jerusalem are reputed to be over 2000 years old, still relative newcomers considering the long domestication of the olive. Man has manipulated the olive tree for so many thousands of years that it is unclear what varieties came from which other varieties. Varieties in one country have been found to be identical to differently named varieties in another. Some research is now being done using gene mapping techniques to figure out the olive family tree. Shrub-like "feral" olives still exist in the Middle East and represent the original stock from which all other olives are descended.

During the Roman Empire some famous experts prepared an olive-oil classification: *Oleum ex albis ulivis* (this could be compared to the extra virgin olive-oil available in these days) and *viride, maturum, caducum, cibarium* (to be considered olive-oils of secondary importance). During the Renaissance period olives received again great importance. With the discovery of America in 1492 olive trees from Europe reached Antille and consequently the American continent. By the 16th century olives are already in Mexico, Peru, California, Chile and Argentina where it is possible to find famous Arauco's olive even today. Recently, olive has started to influence other areas of the World such as South Africa, Angola, Australia, China, and others. Its production extended even to a geographic area with monsoon characteristics that was considered unsuitable few years ago.

In pre-Himalaya regions of Northern Pakistan, Jammu and Kashmir, Himachal Pradesh (H.P.) and Uttar Pradesh of India, and Nepal, olive cultivation are being studied. Some significant results have been obtained for possibilities to exploit olive production in these regions. Olive is known as "Jaitoon" in Sanskrit, Arabic, Hindi, and Nepali. *Olea cuspidata* is a species locally known as "Lotto" in Dolpa and "Launtho" in Bajura districts of Nepal and "Kahu" in Himachal Pradesh, Jammu and Kashmir, and Uttarkhand states of India.

In Pakistan and India the cultivated olives (*Olea europaea* L.) were introduced in fifties as evidenced by the existing plantations in Mingora (SWAT), Rawalpindi, and Pinjore in India, where olive cultivation started 40 years ago. The first olive orchard was established at Jidhari in Solan district by the efforts of Maharaja Patiala's Family. In sixties and seventies, the H.P. government established olive plantations and launched a top-working programme of *Olea europaea* with *Olea cuspidata*. Technical Cooperation Programme of FAO/UNDP was launched to provide necessary technical support from 1975 for several years. With the Bilateral Agreement between Indian and Italian Governments from 1984 to 1993, India has extended commercial olive plantation activities and olive oil industries in Himachal, Jammu and Uttarakhand (Bartolucci and Dhakal, 1999).

References

Anonymous. 2015. The olive oil source.2015.http: //www.oliveoilsource.com/ page/history-olive

Bartolucci, P. and Dhakal, B.R. 1999. Prospects for olive growing in Nepal. *Authors: Pietro Bartolucci and Buddhi Raj Dhakal Technical Editor: Gyan Kumar Shrestha.*

http: //www.livescience.com/26887-olive-tree-origins.html

2

Distribution, Area, Production and Trade

Olive cultivation is flourished since time immemorial and covered significant area under cultivation in areas experiencing wet-cold winters and hot dry summers. Mediterranean climate favours its successful cultivation. However, it has been introduced in newer areas around the world.

Geographical Distribution

The Mediterranean basin is the major region where olive is mainly distributed since ancient times. It was thoroughly distributed by the Romans and later by the Arabs. It was later carried away by explorers and colonists from Mediterranean homeland to England and United States, where it was failed due to unsuitable climate. A long dry season with low humidity is essential to minimize disease, and rainfall during bloom reduces fruit sets. In tropical regions near to equator, olive grow vegetatively, but most cultivars do not set fruits due to insufficient winter chilling, which prevents flower formation. Olive production has been commercially successful in other Mediterranean climates around the world including parts of South America, South Africa and Australia. Therefore today also, majority of olive producing countries are lies in the Mediterranean basin and contributes around 98 per cent of the world total production of olive and olive products. The olive tree is best suited to areas with a Mediterranean climate; a long, hot growing season and a relatively cool winter with minimum temperatures above a lethal limit. The trees do not usually survive below about (-) 12°C and most cultivars are injured at (-) 9°C. Therefore, the commercial olive producing areas of the world are found between 30° and 45° north and south latitudes.

Area and Production

Majority of the olive producing countries are lies in the Mediterranean basin and contributes around 98 per cent of the world total production of olive and their by-products (Figure 1). Presently it is being grown over an area of about 10.2 m ha in the world with a production of 20.34 m tonnes and a productivity of 2. 0 t/ha. Spain is a leading producer with 7.88 m tonnes from an area of 2.50 million hectares and a productivity of 2.69 t/ha followed by Italy, Greece and Turkey (Table 1). Egypt, Syria, Tunisia, Morocco Algeria, Portugal, Lebanon, Israel *etc.* are the other minor olive producing nations, but the productivity however is highest in Egypt (9.79 t/ha) followed by Spain (3.15 t/ha).

Table 1: Major Olive Producing Countries (Year 2013 per FAOSTAT)

Rank	*Country/ Region*	*Production (in tons)*	*Cultivated Area (in hectares)*	*Yield (q/Ha)*
—	World	20,344,343	10,244,194	19.86
1	Spain	7,875,800	2,500,400	31.50
2	Italy	3,022,886	1,125,420	26.87
3	Greece	2,000,000	930,000	21.51
4	Turkey	1,676,000	825,830	20.30
5	Morocco	1,181,675	922,235	12.81
6	Syria	1,000,000	690,490	14.50
7	Algeria	395,776	330,000	11.97
8	Tunisia	963,000	1,800,000	5.35
9	Egypt	510,000	52,100	97.89
10	Portugal	350,900	347,300	10.10

In Asia, cultivation is mostly confined to Iraq, Iran and China however, in India inspite of its vast potential it is grown only in an area of about 400 ha mostly in the Himalayan mountainous region encompassing the three northern states of Jammu and Kashmir, Himachal Pradesh and Uttarakhand hills at an altitude ranging from 1000 to 1300 m above mean sea level. Among the states Jammu and Kashmir leads with an area of 276 ha spread in the districts of Doda, Udhampur, Rajouri, Poonch, Kupwara, and Baramulla. The district Doda has the maximum area (59.76 per cent) followed by Udhampur (15.5 per cent). However, Indian sub-temperate region has great potential can expect 500 (000 tonn) production and 5 ton/ha productivity from (1.0 lakh ha).

Recently, Rajasthan Government has started a ambitious project on olive cultivation in Thar desert of Rajasthan in collaboration with Isreal government. About 112000 saplings of seven varities were imported and planted at Bikaner, Sriganaganagar, Jaipur, Nagour, Jhunjhunu, Jalore and Alwar districts in 182 hectares. The tree started bearing in 2012 with a total production of 10 tonnes and the oil recovery was reported 9-14 per cent.

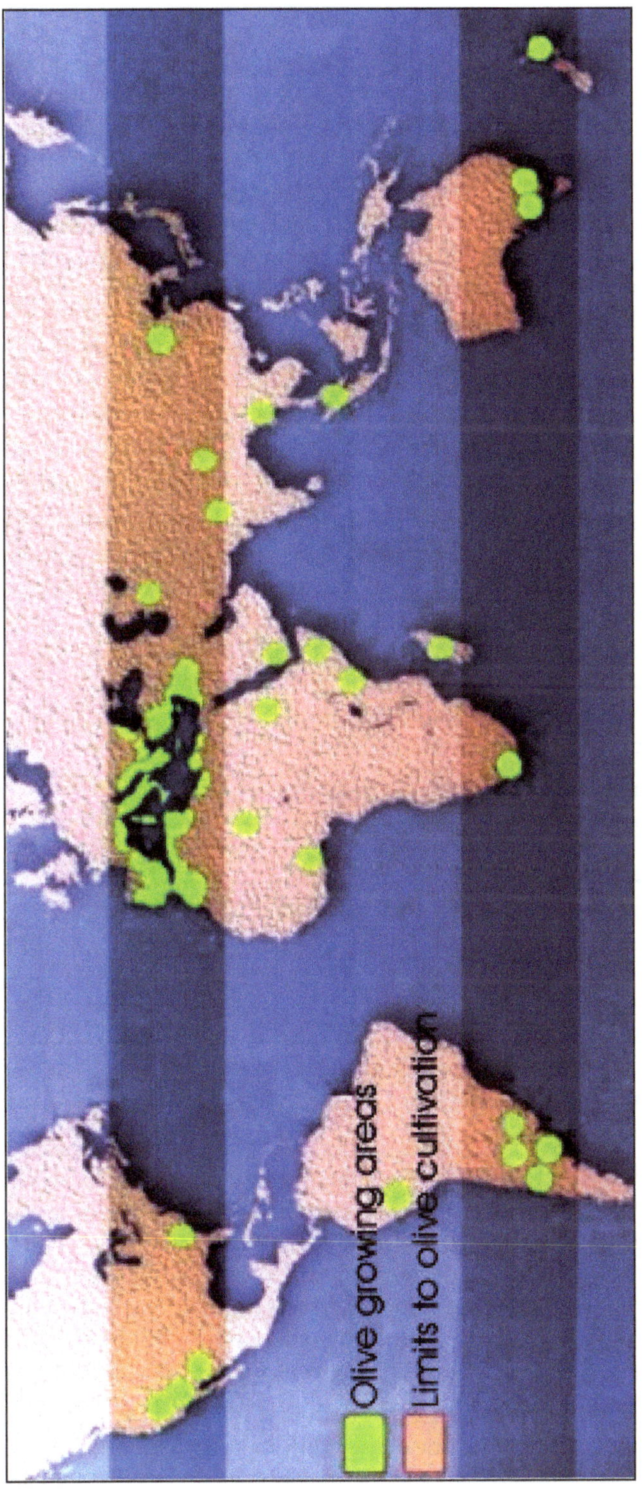

Figure 1: Olive Trees Distribution in the World.

***Source:* International Olive Council.**

Trade

The olive oil market is very complex: production is spread over developed and developing countries and is realized through very different production systems, even within a single country. According to the International Olive Council (IOC) olive oil production is expected to reach 3.098 million tons worldwide, up from 2.425 million tons in (2012/2013). Consumption is also expected to mirror this upward trend, with consumption estimated to reach 3.057 million tons for 2013/2014, up from 3.041 in 2012/2013. Two thirds of EU production is traded internationally (within and outside the EU). Trade within the EU is considerable and continues to rise steadily. In 2010/11 it was around 1 000 000 t, *i.e.* 45 per cent of EU production. Spain is the biggest supplier with 655 000 t, while Italy is the biggest buyer with 533 000 t. EU exports represent approximately 66 per cent of world exports. In 2010/11, exports to third countries amounted to 447 000 t, of which Spain sold 225 000 t and Italy 160 000 t. The biggest markets are the USA, Brazil, Japan, Australia, Russia and China.

In 2010/11, imports accounted for 115 000 t, of which the majority is traditionally under inward processing rules and the remainder within the framework of tariff free quotas with the Mediterranean countries, primarily Tunisia. The new agreement with Morocco has fully liberalised imports from this country. In recent years, India's imports for olive are significant being traded as olive residues, olive oil, olives, and olive preserves (Table 2). In India, according to FAO Database (2012) olive oil is mainly being imported from Spain (46 per cent), Italy (344 per cent), Turkey (3 per cent), Syria (2 per cent), Greece (1 per cent) and Tunisia (1 per cent).

Table 2: Export and Import Scenario of India (FAO, 2011)

Commodity	*Import Quantity (tonnes)*	*Import Value (1000 $)*	*Export Quantity (tonnes)*	*Export Value (1000 $)*
Oil, olive residues	156	398	10	81
Oil, olive, virgin	5910	21543	196	461
Olives	124	80	125	68
Olives preserved	574	1467	15	24

Scope and Potential of Olive Cultivation in India

According to Indian Olive Association (IOA), it is being reported that the Indians are sufferings with lifestyle disorders mainly related to: *Heart Risk*- 10 per cent population affected with cardiac disorders and 50 million peoples suffer from heart problems; *Obesity*- 31 per cent of urban Indians are either overweight or obese; *Stress/Hypertension/BP/Lipids* - hypertension and stress, especially from work, 100 million people have high blood pressure, two out of three employees are a victim of stress and over 40 per cent of urban Indians have high lipid levels (cholesterol and triglycerides) that are the major risk factors for heart disease.

These threats are the major cause of death and involving a major expenditure from hard earned income. Olive is one of the good source provides some kind escape

Table 3: Olive Oil Import Data (FY 2014-15)

Commodity (MT)	*FY 2013-14 (Apr-Mar)*	*FY 2014-15 (Apr-Mar)*	*FY 2014-15 (Apr-June)*	*FY 2014-15 (July-Sept)*	*FY 2014-15 (Oct-Dec)*	*FY 2014-15 (Jan-Mar)*	*Country-wise Import Data (FY 2014-15, Apr-Mar)*							
							Spain	*Italy*	*Tunisia*	*Turkey*	*Greece*	*Portugal*	*USA*	*UAE*
Olive oil virgin	1,689.03	1,962.43	403.63	633.64	500.06	425.10	1,394.62	516.68	4.39	6.73	5.08	0.57	15.04	1.68
Olive oil and its fractions (excluding virgin) of edible	6,060.59	6,564.92	1962.32	1,643.75	1,575.34	1,383.51	4,438.51	1943.25	9.39	117.46	26.69	21.69	2.53	0.28
Other olive oil and its fractions (excluding virgin)	1,691.69	804.60	256.16	238.06	174.62	135.76	420.88	344.24	0.05	1.08	24.8	0	2.7	0.92
Other oil (excluding crude oil) of edible grade not chemically modified from olives	901.28	3,259.33	731.47	1,299.05	232.16	996.65	1,685.44	1018.38	552	3.37	0	0	0.07	0
Other oil other than edible grade (excluding crude oil) from olives	375.74	29.37	2.5	5.51	0.98	20.38	—	28.61	0	0	0	0	0.76	0
Total	**10718.33**	**12,620.65**	**3,356.08**	**3,820.01**	**2,483.16**	**2,961.40**	**7,939.45**	**3851.16**	**565.83**	**128.64**	**56.57**	**22.26**	**21.1**	**2.88**
Share (per cent)							62.91	30.51	4.48	1.02	0.45	0.18	0.17	0.02

Source: Ministry of Commerce and Industry, Govt. of India.

mechanism and avoids the health hazards. The crop is mainly grown for extraction of oil as well as for table purposes and pickle making. In India, the demand of olive is increasing very fast due to its peculiar medicinal properties. Keeping in view the medicinal value of the olive oil, its consumption is increasing many folds. As a result, its imports in terms of value has increased from Rs. 52 Crore (2006) to Rs. 550 Crore by the end of 2012 and expected to increase by Rs. 1000 Crore by 2020. Therefore, there are greater opportunities of growing olive in India.

References

FAOSTAT.2011. http: //faostat.fao.org/site/567/DesktopDefault.aspx?PageID=567#ancor

FAOSTAT.2013. http: //faostat.fao.org/site/567/DesktopDefault.aspx?PageID=567#ancor

IOA.2014. Indian olive association. http: //www.indolive.org/Import-data-2014-15.pdf

IOC.2015. International olive council. http: //www.internationaloliveoil.org/projects/paginas/Section-a.htm

3
Taxonomic Classification and Species

Olea europaea, the olive tree, means olea-olive and europaea-Europe. So *Olea europaea* is the "olive from Europe". The common name olive comes from the Latin word *oliva* and the Greek word *elaia*. Olive belongs to the family Oleaceae, genus *Olea*. It has two main subspeces *i.e. Olea europaea* ssp. *cuspidata* (Wall. ex G. Don) Cif. – African olive; *Olea europaea* ssp. *europaea* L. – European olive. African olive has several synonyms includes Synonym(s): *Olea africana* Mill., *Olea chrysophylla* Lam., *Olea europaea* ssp. *africana* (Mill.) P. Green, *Olea ferruginea* Royle, and *Olea verrucosa* (Willd.) Link. Therefore, it can be placed as under

Kingdom	:	Plantae
Phylum	:	Anthophyta
Class	:	Dicotyledones
Order	:	Scorphulariales or Lamiales
Family	:	Oleaceae
Genus	:	Olea
Species	:	*Olea brachiata*
		Olea capensis
		Olea caudatilimba
		Olea chryssophylla
		Olea europaea
		Olea exasperate
		Olea guangxiensis
		Olea hainanesis

Olea laurifolia

Olea laxiflora

Olea nerifolia

Olea oleaster

Olea oleaster

Olea paniculata

Olea parvilimba

Olea rosea

Olea salicifolia

Olea sylvestris

Olea tetragonoclada

Olea tsoongii

Olea undulate

Olea woodiana

Bernard *et al.* (2008) developed a phylogenetic tree based on the genome content estimated by flow cytometry. They clearly demarcated the importance between cultivated and wild species. The *Olea europaea* subsp. *europaea* is the primary cultivated olive subspecies grown in Mediterranean region. Five more subspecies also described which are non-Mediterranean type are subspecies *laperrinei* (Saharan massifs), *cuspidata* (South Africa to South Egypt and from Arabia to North India and South-West China), *guanchica* (Canary Islands), *maroccana* (Agadir Mountains, Morocco) and *cerasiformis* (Madeira). In recent phylogeographic studies revelled that *O. europaea* subsp. *cuspidata* has diverged early from North African and Mediterranean taxa. A high diversity was also reported in North-West Africa where four subspecies occur (subsp. *europaea, maroccana, guanchica* and *cerasiformis*), because of geographic isolation and fragmentation of habitat in this area.

There are about 600 species and 24 genera reported in Oleaceae. Among these 33 shrub species and 9 tree subspecies of evergreen in nature. In addition, these taxa are classified in three subgenera, *Olea*, *Paniculatae* and *Tetrapilus. The* subgenera *Olea* again have been subdivided into two sections *i.e.*, *Olea* and *Ligustroides*. The section *Olea* is mainly comprised of *Olea europaea* subspecies. This subgenus *Olea* is mainly distributed from South Africa to China, across the Saharan mountains, Macaronesia and the Mediterranean basin. However, the *O. europaea* species is also found outside of its native range as a result of human-mediated dispersal; it has been repeatedly introduced in the New World including Australia, New Zealand and the Pacific islands. *Olea* taxa are found from subtropical and tropical areas. Some of the taxa of subgenus *Olea* occur in arid environments. However, majority of the olive complex (section *Olea*) is found in open forests of the Mediterranean and subtropical regions of the Old World, from Macaronesian cloud forests (subspp. *cerasiformis* and*guanchica*) to extremely arid Saharan mountains (subsp. *laperrinei*). Based on biochemical and

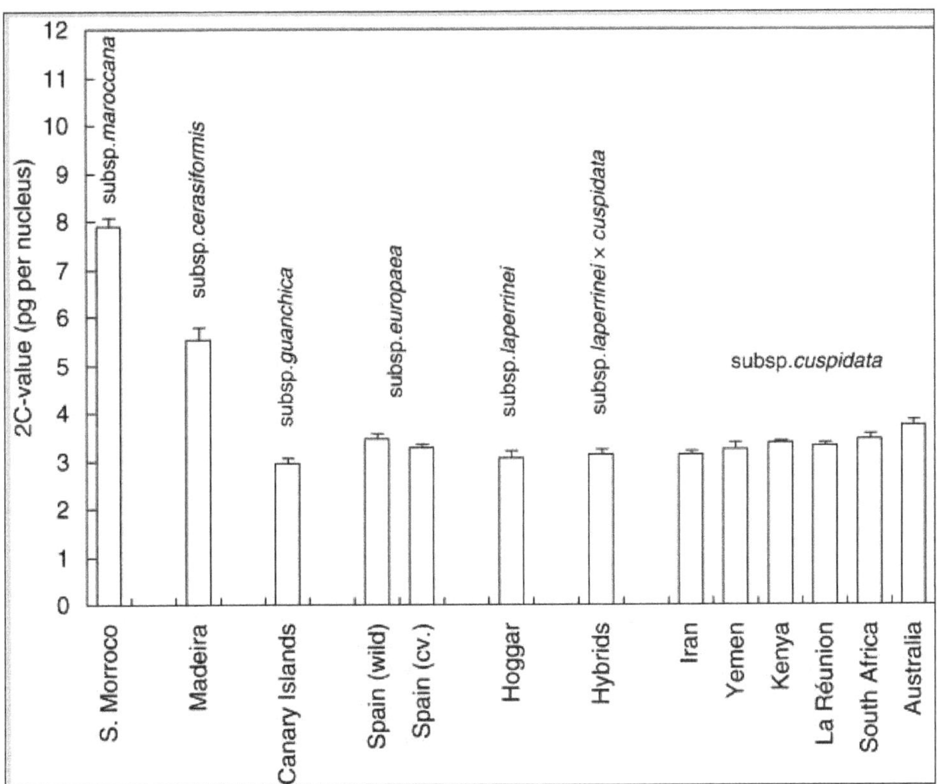

Figure 2: Phylogenetic Diversity Based on Genome Content (in pg per nucleus) Measured through Flow Cytometry in Olive (*Source*: Besnard *et al.*, 2008).

molecular markers, Olea species have grouped in different categories. A phylogeny for Olea has been suggested based on plastid DNA sequences and nuclear ribosomal DNA (nrDNA) restriction fragment-length polymorphisms (RFLPs), respectively. However, a little is known about the biogeography and evolutionary history of other *Olea* taxa, but a better understanding of *Olea* taxonomy and diversification may be of great importance for the future management of olive genetic resources and for the *in situ* conservation of genetically differentiated entities.

Wallander and Albert (2000) divided Oleaceae into five tribes, Oleeae, Jasmineae, Myxopyreae, Forsythieae and Fontanesieae. The *Syringa* and *Ligustrum* form a clade within the tribe Oleeae. The tribe Jasmineae is sister to Oleeae; the remaining tribes (Fontanesieae, Forsythieae, Myxopyreae) are more distantly-related and basal within the Oleaceae.

Species

Oliaceae has more than 30 species native to South Africa, New Zealand, India and Afghanistan. The species *Olea chrysophylla* is considered to be its wild ancestor who is native to Africa and Asia. It is highly polymorphic species as nine different geographic types have been described; some of them such as, *O. excelsa* of the

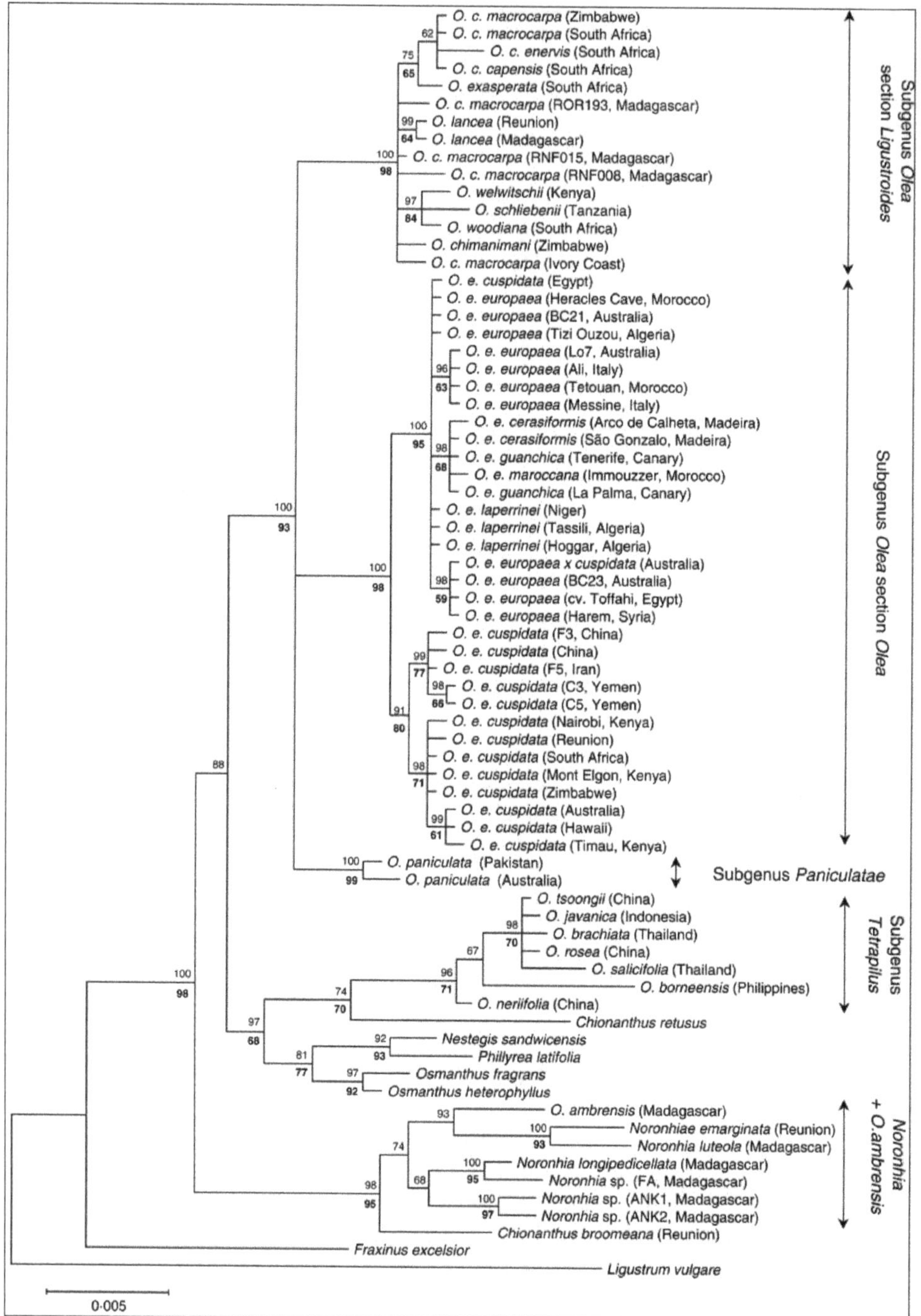

Figure 3: Phylogenetic Tree for *Olea* Species Based on Four Plastid DNA Regions (*trnT-trnL*, *trnL-trnF*, *trnS-trnG* and *matK*).

Abbreviations*: *O.c., Olea capensis*; *O.e., Olea europaea.

(*Source:* Besnard *et al.*, 2009).

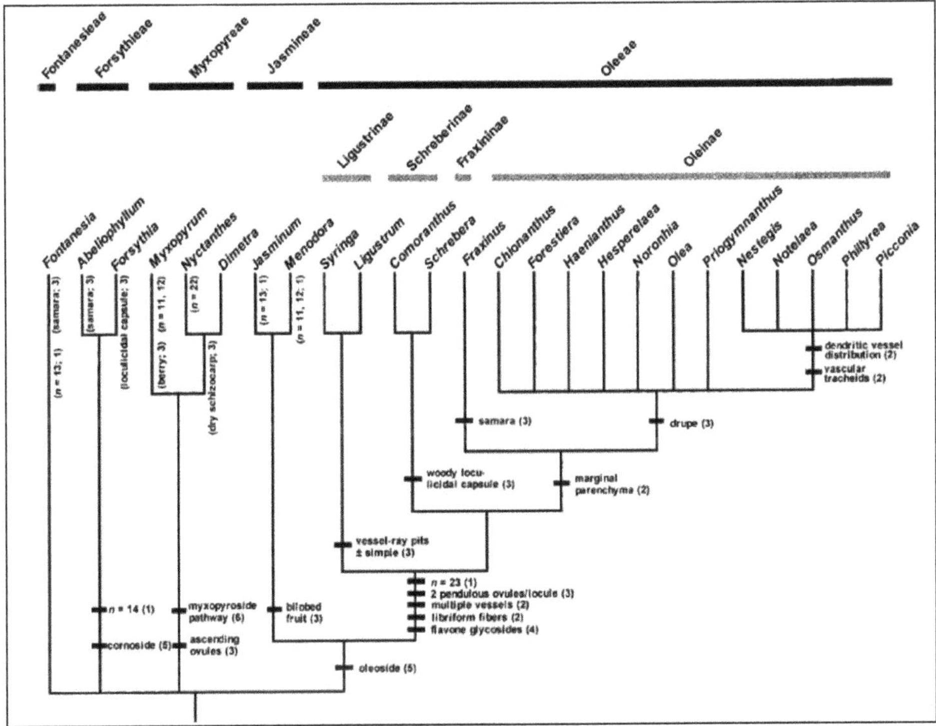

Figure 4: Molecular Phylogeny of the Oleaceae. The dark bar at the top delimits the various tribes, the grey bar beneath, the subtribes of the Oleaceae.

(*Source*: Wallander and Albert, 2000).

Canary Islands have been evaluated to the status of the species. Intermediate taxa, between *O. chrysophylla* and *O. europaea*, is *O. laperrini* found in central Sahara and in Morocco. Some, however, consider the Euro-Mediterranean olive (cultivated *sativa* or wild *oleaster*), the *laperrini* and the Asiatic cuspidate as the three species of *O. europaea*. However, there are at least five natural subspecies distributed over a wide range: *Olea europaea* subsp. *europaea* (Europe), *Olea europaea* subsp. *cuspidata* (Iran to China), *Olea europaea* subsp. *guanchica* (Canaries), *Olea europaea* subsp. *maroccana* (Morocco), *Olea europaea* subsp. *laperrinei* (Algeria, Sudan, and Niger). The salient characteristics of these species are given below:

Olea chrysophlla

It is native to North Africa, however, found in abundance in Asia also. The species is highly polymorphic and numbers of its geographic types are common in its native place. Small tree, branches are slender or scurfy upward. Leaves are lanceolate, 5-10 cm long, petiole short or very short. Flowers are small in size (0.3 cm long) and are borne in axillary panicles. Fruit is a drupe, globose or somewhat long in shape and blackish in colour.

Olea cuspidate (Indian Olive)

The plants of this species are found in abundance in the foothills of the Himalayas from Kashmir to Kumaon up to 2400 m altitude. This species is closely related to *O. europaea* medium sized tree. Leaves are oblong-lanceolate. Flowers bisexual, whitish. Fruit is a drupe, ovoid 5-8 cm long, black when ripe, edible but the pulp is scanty and not particularly pleasant in taste. Both pulp and seed yield oil (Srivastava and Kapoor, 1985).

Olea dioca Roxb.

Distributed in Eastern Himalayas, Duars, Assam and in Deccan peninsula chiefly in Western Ghats. Its wood has sweet scent thus known as Rose sandalwood, suitable for carving and cabinet work, also reported for medicinal uses. The bark is reported to be used as febrifuge. The leaves may be used as green manure.

Olea europaea var. *communis* (Cultivated Olive)

A handsome, grey-green tree; bark is grey in colour and branches are obtuse angled. Leaves are lanceolate, willow-like in shape and fruits are globular or oblong.

Olea europaea subsp. *europaea* var. *europaea*

Commercial olives are being produced from this species are of edible quality. Trees are small trees rarely exceed 6-8 m a few can up to 20 m. The wood is resistant to decay, and when the top of the tree is killed by mechanical damage or environmental extremes, new growth arises from the root system. The root system is generally is shallow, spreading to 0.9 - 1.2 m even in deep soils. The above ground portion of the olive tree is recognizable by the dense assembly of limbs, short internodes, and compact nature of the foliage. Leaves opposite thick, leathery, growing over 2-years period. Stomata on abaxial surface nestled in peltate trichomes reducing transpirational loss of water, which imparts relative drought resistant to olive tree. Leaves lanceolate or oblong in shape. Racemes of yellowish white fragrant flowers are borne on forking panicles emerging out from leaf axil, considered a native of Asia Minor. Fruit attains horticultural maturity during November.

Olea ferruginea Royle syn *O.cuspidata* Wall. ex G. Don

It is found in Himalayas from Kashmir to Kumaon hill and known as Indian olive; wood is whitish, it's durable and resistant to fungal attack. Fruits are edible. The timber is used chiefly for tool-handles, walking sticks, croquet balls, mallets, combs, toys, turnery articles, carving, ploughs, and ginning machines and boat-building. Suitable for engraving, printing blocks and for mathematical instruments. The fruits are edible. Leaves and bark are used as antipyretic in fever and debility. Leaves considered as a cure for gonorrhoea.

Olea grandulifera Wall. Ex G. Don

In India it is found in outer Himalayas from Kashmir to Nepal, hills of South India. A handsome, medium -sized the wood is used for house construction, carpentry, agricultural implements and turnery. The bark and leaves are astringent and are used as antipyretic in fevers. Plants small bushy glabrous tree, bark is dull grey in colour, branches spreading and alternate. Leaves are large in size, elliptic-lanceolate in shape. Fruits are small in size, round to oblong and inedible. Flower white bisexual in terminal and lateral compound pyramidal panicle, petal rounded.

Olea laperrini

Widely spread in Morocco and Central Sahara. Is regarded as an intermediate taxa between *Olea europaea* and *Olea chrysophlla.*

Olea europaea var. *oleaster* or *Olea europaea* var. *sylvestris* (Wild Olive)

Plants are medium to tall, 20 feet sometime even more, wild form, thorny, branches are four angled. Leaves are elliptic or oblong, fruits small, roundish and inedible. It is used as rootstock for cultivated forms.

Distant Cousins

Russian Olive or Oleaster

The Oeaster (*Elaeagnus Angustifolia*) belongs to the family Elaeagnaceae and not used as fresh olives or olive oil and distantly related to the olive tree. They share the same class, Magnoliopsida (Dicotyledons) but different order, species, *etc.* There is evidence of cross-reactivity between olive, ash, privet, and Russian olive tree pollen allergens. Russian olive is a native of southern Europe and western Asia. It was introduced into the United States in the early 1900's and has now escaped cultivation and is extensively naturalized in 17 western states. Once established, Russian olive is hard to control and nearly impossible to eradicate. Control efforts have included mowing, cutting, burning, spraying, girdling, and bulldozing, most with limited success. It does produce a small fruit that is nutritious to deer, cattle, birds and rodents but when Russian olive displaces natural species the resultant habitat is generally considered inferior.

Figure 5: Russian Olive.

Sweet Olive

It is *Osmanthus Fragrans*, also known as tea olive, fragrant olive belong to family (Oleacea). Other Genera in the family are: Ligustrum (Privet), Syringa (lilac),

Fraxinus (ash) Synonyms: *Olea fragrans, Olea ovalis, Osmanthus* longibracteatus, *Osmanthus macrocarpus*. The sweet olive does not make a seed with appreciable oil and is not used for making oil. The olive which is delayed harvest has a sweeter and more ripe flavor is sometimes referred to as "sweet olive oil" on restaurant menus, as opposed to an astringent, bitter Tuscan style oil. Bland refined olive oil is also referred to as "sweet olive oil" when it is sold in drug stores for softening earwax, *etc.*

References

Besnard,G., Garcia-Verdugo, C., Rubio De Casas, R., Treier, U.A., Galland, N. and Vargas, P. 2008. Polyploidy in the olive complex (*Olea europaea*): Evidence from flow cytometry and nuclear microsatellite analyses. *Ann. Bot.* **101**(1): 25–30.

Besnard, G., Rubio de Casas, R., Christin, P.A. and Vargas, P. 2009. Phylogenetics of Olea (Oleaceae) based on plastid and nuclear ribosomal DNA sequences: Tertiary climatic shifts and lineage differentiation times. *Ann. Bot.* **104** (1): 143-160.

Srivastava, S. K. and Kapoor, S. L. 1985. A taxonomic revision of Indian Olea Linn. (Oleaceae). *J. Econ. Taxon. Bot.* **7**(1): 167-177.

Wallander, E. and Albert, V.A. 2000. Phylogeny and classification of Oleaceae based on rps16 and trnL-F sequence data. *American Journal of Botany* **87**(12): 1827-1841.

4
Botany

Olive is member of family *Oleaceae.* The species *Olea eurọpea* is a only species which produces edible fruits. *Olea cuspidata* also known as Indian olive are found in the Himalaya from Kashmir to Kumaon upto 2400 m altitude and this species is closely related to *Olea europea* and mostly used as root stock. The other species are *Olea crysophylla, Olea verrucosa* and *Olea laperrini* all are native to Africa but bear no edible fruits.

The place, time and immediate ancestry of the cultivated olive are unknown. It is assumed that *Olea europaea* may have arisen from *O. chrysophylla* in northern tropical Africa and that it was introduced into the countries of the Mediterranean Basin via Egypt and then Crete or the Levant, Syria and Asia Minor. Fossil Olea pollen has been found in Macedonia, Greece, and other places around the Mediterranean, indicating that this genus is an original element of the Mediterranean flora. Fossilized leaves of Olea were found in the palaeosols of the volcanic Greek island of Santorini (Thera) and were dated about 37,000 Before Present (BP). Imprints of larvae of olive whitefly *Aleurolobus* (*Aleurodes*) *olivinus* were found on the leaves. The same insect is commonly found today on olive leaves, showing that the plant-animal co-evolutionary relations have not changed since that time.

Olive is an evergreen tree. Leaves are replaced every after 2–3 years. Tree height varies from 5 to 20 m. The plant has either spreading nature or bushy habit in its natural state (basiplastia); but can be trained to a tree. Root system is strong; distributed almost all over top soil but with roots having capability to hint rocky soil. Horizontal growth of roots is almost three times the canopy width. The trunk is cylindrical, tortuous, sometimes with characteristic knot-like swellings; the trunk diameter can be more than two meter with a light gray bark. It is a long-lived tree; some trees have been lived for 1000 years. The olive wood is very hard and delicately grained. Leaves are lanceolate, thick, leathery and opposite. Leaf color is green-grey on the upper surface and silver-green on the lower surface. Leaves have stomata nestled in peltate trichomes on their lower surface (star-like hairs) that restrict water loss due to transpiration and make the olive relatively resistant to drought.

Inflorescence

A mature olive (*Olea europaea* L.) tree produces about 500,000 flowers but only 1–2 per cent of them set fruit that reach maturity (Griggs *et al.*, 1975, Martin, 1990 and Lavee *et al.*, 1996). Olive flowers are borne on inflorescences which called as panicles. The inflorescences are mostly developed at leaf axils and have a central axis which is terminated by a flower. The primary branches arise from the central axis and may also have secondary branches. In some cultivars, tertiary branches are also found. The number of flowers and their distribution on the inflorescence are specific for each cultivar but can change from year to year (Lavee, 1996).

Flower

The olive bears small and white flower. It comprised of four fused green sepals, four white petals, two stamens each with a large yellow anther and two carpels each with two ovules. Flowers are either perfect (hermaphrodite) or staminate (male). Perfect flowers characterized by a plump green pistil with a short, thick style and a large stigma. Staminate flowers do not have pistil or only a tiny, yellow aborted one (Lavee *et al.*, 1999). The proportion of perfect and staminate flowers varied according to genetic, climatic conditions, and the level of fruit production in the last year; therefore, it may vary from season to season, from tree to tree, from branch to branch, and from inflorescence to inflorescence. Both the viable pollen grains produced by both perfect and staminate flowers but only the perfect flowers set fruit (Arzee, 1953).

Fruit

The fruit of the olive tree is a drupe which comprised of the epicarp or epidermis, the mesocarp or flesh and the endocarp or pit which consists of a woody shell enclosing one or, rarely, two almonds (seeds). However, it differ from all others drupes in their chemical composition as its fruits have low concentration of sugars, 2-5 per cent versus around 12 per cent, a high oil content, 20-30 per cent versus 1-2 per cent, and in their characteristic strong bitter taste. The strong bitter taste is due to the presence in the olive of the glucoside, oleuropein, which does not occur in any other fruit or tissue in the plant kingdom.

Seed

The olive seed developed by in various stages. The fruits of olive horticuturally mature in Indian condition in the month of mid September to mid October and November and physiologically in the month of November. The mature seed is covered with a thin coat that covers the scratch-filed endosperm. The latter surrounds the tapering, flat, leaf like cotyledons, short radical (root) and plumule (stem). Seed size and absolute shape vary greatly according to cultivars.

Root System

The root system of olive is featured by shallow root systems. This allows olive roots to collect water from soil that have an attributes of drying fast. The shallow root system ensures the tree gets enough moisture to stay hydrated. While shallow root systems get water faster, they can also become exposed more easily. There is always possibility of damage from the elements and from landscaping tools such as lawn mowers and weed trimmers.

Leaves

In olive (*Olea europaea* L.) leaf is simple, subsessile, thick, coriaceus, lanceolate to obovate, more or less pointed at the apex. The margins are entire and folded abaxially. Nervation of the leaf is reticular. The upper surface is dark-green, glabrous and shiny, the lower surface shimmering silver, and tomentous (densely pilous) Branislava *et al.* (2007).

Main Branches

The branches orginiate at a height of 1.2 m in the classical olive grove and at 20-40 cm in the modern dense olive plantings. The number of branches is three or more. The main branches give secondary and territory branching bearing the leaves, flowers and fruits. The small shoots are classified in to four categories:

1. *Vegetative shoots*: Bearing only vegetative buds and producing new shoots and leaves.
2. *Fruit bearing shoots*: Bearing flower buds: their number is greater in the low vigor trees.
3. *Mixed shoots*: Bears vegetative and flowering buds concurrently. The flowers and fruits are borne at the base of the mixed shoots.
4. *Water sprouts*: Originating from the trunk, branches and the thick shoots. These are very vigorous, grow vertically and they should be removed unless they are going to substitute for a low vigour branches.

Trunk

The olive trunk is cylindrical, with an uneven surface, bearing a lot of swellings. The wood is yellowish and darker towards the centre of the trunk.

References

Arzee, T. 1953. Morphology and ontogeny of foliar sclerids in *Olea europaea* L. I. Distribution and structure. *Am. J. Bot.* **40**(9): 680-687.

Branislava, L., Vloletapo, P. and Dušanka R. A. 2007. Morpho-anatomical characteristics of the raw material of the herbal drug olivae folium and its counterfeits. *Arch. Biol. Sci., Belgrade* **59** (3): 187-192.

Griggs, W.H., Hartmann, H.T., Bradley, M.V., Iwakiri, B.T., Whisler, J.E. 1975. Olive pollination in California. *California Agricultural Experimental Station, Bulletin* 869.

Lavee, S., Rallo, L, Rapoport, H.F., Troncoso, A. 1996. The floral biology of the olive: Effect of flower number, type and distribution on fruitset. *Scientia Horticulturae* **66**: 149-158.

Lavee, S., Rallo, L., Rapoport, H. F. and Troncoso, A. 1999. The floral biology of the olive II. The effect of inflorescence load and distribution per shoot on fruit set and load. *Scientia Horticulturae* Amsterdam **82**(3-4): 181-192.

Martin, G.C. 1994. Botany of the olive. In: Ferguson, L., Sibbett, S.G., Martin, G.C. eds. Olive production manual. University of California, Division of Agriculture and Natural Resources, Oakland, CA, California, pp. 19-21.

5

Floral and Fruit Biology

Flowering is a critical step in fructification. No flowers mean no fruit, and subsequently low crop load may be limited by the number of flowers formed. As in some of the fruit species which form a large amount of flowers, fruit set rather than flower number is the parameter which usually determines yield. Apart from their importance in the determination of crop yield, some events occurring during flower formation and set affect fruit let development and final fruit size and quality, having an additional effect on returns. The study of these processes has not only an academicals interest but also an applied aspect. The information on the regulation of flowering and fruit set is very advantageous in all fruit crops and can be manipulated to the advantage of the grower as well as markets.

Olive tree life cycle can be divided in to four stages:

1. **Unproductive stage**: During this stage the tree grows at high rates and is characterized by the absence of flowering and fruiting. Lack of production is due to the absence of a sufficient equilibrium between the canopy and the root system.
2. **Stage of increasing production**: Flowering means production and the tree increase its productive capacity as time passes. Its canopy grows and with it the number of buds that are susceptible to flower induction.
3. **Maturity stage**: During which plant size and production have attained a maximum. Productivity can be considered constant although it may fluctuate greatly from year to year. In this stage the olive grove produces its best.
4. **Senescence stage**: All processes typical of aging (low vegetative, reduction of expansion of root system, abundant flowering followed by poor fruit set, susceptibility to disease) appear and indicate a tendency of the tree to weaken and die.

Olive Annual Biological Cycle

- ☆ Rest period
- ☆ Period of active vegetative growth
- ☆ Flower bud differentiation
- ☆ Anthesis-fruit set
- ☆ Fruit growth
- ☆ Pit hardening
- ☆ Ripening
- ☆ Harvesting
- ☆ Pruning

Flower Induction

The exact timing of flower bud induction is not known; however, most of the reports suggested that induction occurs between July and February, depending on cultivar and environmental conditions. Consequently, flowers buds are developed one year before they set a crop. Flower bud inflorescence is borne in the axil of each leaf. Usually, the bud is formed on the current season's growth and gives visible growth in the next season. There are hundreds of flowers per twig. Each inflorescence contains 15 to 30 flower buds which are small, round and white yellow in color. The flowers are white or whitish; the calix is short and 4 - toothed; the corolla is short-tubed with 4 valvate petals; the stamens are 2, each bearing 10.000 to 15.000 small and light pollen grains. The ovary is 2- loculed, bearing a short style and a capitate stigma. The pistil has two carpels, each containing two ovules but only one is fertilized and thus produces one-seeded drupe. There are two types of flowers: perfect flowers that contain stamen and pistil, while the staminate flower contains aborted pistil and functional stamens. Large commercial crops occur when 1 to 2 perfect flowers are present among 15 to 30 flowers per inflorescence. The perfect flower has large pistil which almost fills the space within the floral tube. The pistil is green when immature and deep green at the time of full bloom. Olives are monoecious; the same tree bears perfect and imperfect flowers. The reasons for flower and young fruit drop are not well known. However, pistil abortion is often involved. Stress from lack of water and nutrients during floral development can lead to pistil abortion.

Flower buds begin to form by winter and full bloom occurs during April -June depending upon the cultivar and the latitude under Indian conditions. Flower drop can also be due to (a) a sudden decreasing of temperature, (b) continuous rainfall, (c) thick fog, (d) strong winds, (e) too low Relative Humidity, and (f) nutrient deficiency The biological fruiting cycle starts with the flower bud differentiation (Feb.-March) and ends with fruit ripening (Nov.-Dec.) *i.e.* it takes about 9 months duration. The main phonological phases are (a) flower bud differentiation, (b) panicle/raceme formation, (c) blooming, (d) pollination, (e) fertilization, and (f) fruit set and fruit development.

Flower Bud Differentiation

Understanding the flower bud differentiation in olive trees is very important to regulate annual production to overcome alternate bearing habit which is characterised by almost no vegetative growth with profuse flowering and heavy fruiting ('on' year), leading to a shortage of flowering sites on the branches, which hold the potential for next year's crop ('off' year) (Gold-Schmidt, 2005). Flower bud formation requires a series of changes in the differentiation pattern of apical or axillary buds. Flower bud development is a highly complex process that is characterised by two distinct physiological phases: (a) bud initiation and (b) floral bud development (Bernier, 1988). Internal and external signals have a fine regulatory function with regard to the timing of flower-ing (Fernandez-Escobar *et al.*, 1992) in olive, and it was reported that flower induction occurs in February and March, approximately two months before flowering (Fab-bri and Alerci, 1999; Hartmann, 1951; Monselise and Goldschmidt, 1982).

The time of bud formation is neutral; they can develop into vegetative buds (shoot) or flower buds. Olive tree can have also mixed buds; they produce floral organs and shoots. A floral behaviour study performed in India (Indo- Italian Olive and Fruit Dev. Project) at Basht and Govindpora of Jammu and Kashmir showed that buds, containing the relative primordia of so called "mixed buds" are insignificant in number (*i.e.* less than 1 per cent on the average) (Giannotti, 1992). In the olive, as in all common evergreen plants, the differentiation of buds starts in one-year old twig during the year of flowering. Excessive growth, unfavourable climatic conditions, heavy pruning, *etc.* can reduce the number of differentiated buds.

Panicle/Raceme Formation

Floral organs are ready to open one month after the flower bud appearance. The inflorescence is composed of one main rachis bearing a flower of first order which has several secondary rachis with flowers of second order. There are flowers of third order also. On each rachis or axil, flowers are borne opposite to each other. Some cultivars have only one axel and all flowers of second order are attached to this one. The difference in the inflorescence structure is a typical character of olive cultivars.

Blooming

Blooming starts immediate as floral organs are completely developed and formed and at this stage flowers are ready for pollination and fertilization. Each flower cluster bears 15 to 30 small, roundish and white-yellow blossoms. The flowers are also small, may be imperfect due to pistil hypotrophy and consists of 2 stamens each bearing 10,000 to 15,000 pollen grains. These pollen grains are carried by wind up to a distance of 12 km.

Anthesis, Dehiscence of Anthers and Stigma Receptivity

The peak period of dehiscence of anthers was observed in olive between 02=00 hrs to 06=00 hrs. The anthesis is favoured by low temperature coupled with a moderate relative humidity. The pollen grains released during early morning hours can be conveniently used for pollination immediately after their dehiscence from pollen sacs. Fresh pollen grains appear as a mass of golden yellow colour with naked eye but they appear oblong and pale yellow in colour under a low power microscope. In aqueous medium, pollens attain a circular shape. The extent of pollen germination varied with the cultivar and season. However, a pollen germination of 28.0 to 48.0 per cent has been observed in olive cultivars. Peak period of stigma receptivity varied between one day before and one day after anthesis.

Pollination

Olives produces both hermaphroditic or 'perfect' flowers (containing male and female parts) and staminate flowers (containing male parts only) on the same plant this condition botanically called as andromonoecious. Each inflorescence may contain both perfect and staminate flowers. The proportion of staminate to perfect flowers is determined approximately 4 weeks prior to bloom. One or two perfect flowers within an inflorescence are sufficient to support a commercial crop. During early flower bud development, all buds are perfect. Imperfect flowers result from pistil abortion (loss of female flower part). During bloom one can visually differentiate between perfect and imperfect flowers. The flowers are whitish and small and each bears a 'short-toothed' calyx and 'short-tubed' corolla (Hartmann and Opitz, 1966). Hermaphrodite flowers generally have two stamens and a bi-locular ovary with a short style and stigma (Hartmann and Opitz, 1966). In staminate flowers, the pistil is either rudimentary or absent. The flowers are wind pollinated, bear large quantities of pollen, and no nectar is produced as they do not possess nectaries (Martin, 1994). Griggs *et al.* (1975) observed that olive flowers are morphologically adapted to either self- or cross-pollination. In some flowers, the anthers are close enough to the stigma so that when they dehisce, the pollen falls on the stigma and self-pollination could occur. At the same time there are flowers where the filaments are flattened so that the anthers spread away from the stigma, thus favouring crosspollination. According to Cuevas *et al.* (2001), olives are naturally suited to cross-pollination by wind. The presence of flowers with male parts only indicates that these flowers are formed for the sole purpose of acting as pollen donors. The abundant amount of pollen, up to 200,000 pollen grains per flower, shows that they are adapted to be wind pollinated.

Pollen grains reach the stigma, which is receptive for several days with some exceptions. In cultivar "Luques", the stigma receptivity is of 8 days but the pollen tube growth requires four days; so the total pollination period is only four days. The cultivar "Picholine" has eight days of pollination period. In most of the olive cultivars flowers of one tree/cultivar are pollinated by pollens of other tree/cultivar. There are exceptions; the cultivars such as "Frantoio" and "Ascolana Tenera" have self compatible pollens, but even these cultivars perform better with cross pollination (Nilsson, 1998). It is very important to find proper cultivar for pollination and their placement in the orchard is also important. Pollinators in new olive plantations should be around 20 per cent of the total trees and they are to be located in the field towards the prevailing wind. Cultivar combination may also be important for good production. "Pendolino and Sevilano" are good pollinizers for many cultivars. "Pendolino" is a commercial pollinizer variety for Frantoio and Leccino cultivars.

Flower Fertilization

As soon as fertilization takes place there is a induction of growth of flower tissue for fruit formation. There is also some sterility phenomenon that does not allow fertilization to take place. Fruit formation may take place without fertilization or with partial fertilization. In such cases, fruits are small called "olive passerine or shot berries" other factors of sterility in olives are lack of nutrients and climatic factors.

Fruit Set and Fruit Formation

Fruit set is considered normal when 2 to 4 per cent of flowers converts in to fruits. Fruit formation start with pericarp formation followed by epicarp (skin), mesocarp (pulp) and then endocarp (stone) development. One month after fruit formation, about 20 per cent of fruits drops under normal condition. From September (in Northern Hemisphere) there is a change in the colour of epicure (from green to red-violet, black) and such change indicates ripening process occurring inside the fruit. At Bajaura of Himachal Pradesh, India, Pietro Bartolucci, Gaetano Tassone and one Indian Horticulturist studied fruiting behaviour of 16 cultivars in 1992. They observed that fruit drop was equal to or less than the standard value recorded in common olive growing areas during monsoon. Weekly observation of fruit branches and twigs showed that the average fruit set decreased from 47. 4 per cent (soon after full bloom period) to 2. 9 per cent (at ripening stage).

Fruit Development

Different phases

There are two different mechanisms responsible for increase in size of the drupe. The young drupe accumulates water then cells are swollen simultaneously cells also proliferate rapidly. The first sign in fruit differentiation is noticed by the hardening of the stone due to appearance of sclerenchymous cells. This step is very important about 50 days after fertilization because most nutritional resources are mobilized by this stage and all other parameters slowdown at this stage. Pulp development begins and then oil starts to accumulate. Oil content as oil composition is depends on variety to variety with little influence of the environment whereas other compounds (sterols, phenols,) are much dependent of the environment.

Parthenocarpic Fruits (Shotberry)

Self-pollination and cross-pollination modify the level of seedless fruits bore by most of the varieties (Koubouris *et al.*, 2009). The yield in shotberries is due to self-incompatibility and or inter-incompatibility and thus it signifiaces that pollination is not optimal in the orchard. Some varieties display commonly different fruit size that result from fertilized and shotberries.

Stone and Embryo Number

The olive fruit stone generally contains one embryo (seed), but few may contain no seed and some contains two seeds. The seed number is variable between varieties and may indicate more or less that not enough inter-compatible pollen has landed on the stigma (Farinelli *et al.*, 2006). The embryo number is an important parameter to forecast whether the drupes will fall easily before harvest or not. The more attached drupes contain two embryos. If so, to enhance efficient pollen grain number in a monovarietal orchard could be obtained by enhancement of the adequate pollinizer number.

Storage Compound Accumulation

The different genetic origins of the flesh and the embryo make deep differences in compounds accumulated in these organs. The drupe tissues are of maternal origin and thus they belong to same genetic composition. Oil accumulated in the flesh has the same composition for all drupes of a variety (no genetic variance), but it may be affected by environmental conditions. For phenol compounds the variation is important and is exploited through the appellations. In contrast for the embryo, the genetic variability exists between fruits from two origins: firstly from the allele segregation of the mother plant, each ovule has a specific genetic composition, and second the diversity comes from the pollen grain genome, each pollen grain has also a specific genetic composition (Olliver *et al.*, 2006). If the commercial olive oil should have between 55 per cent to 85 per cent of oleic acid, the oil in the embryo has usually less content in oleic acid (about 30 per cent) and the major fat is the linoleic acid similar as in sunflower oil. However, the fatty acid is not free but each esterifies a alcohol radical from glycerol and the different isomers are used to recognize appellations (Olliver *et al.*, 2006).

Alternate Bearing

Alternate bearing, described as a sequence of heavy yields followed by light ones over several years, affects several temperate, tropical and sub-tropical fruit tree species such as apple, pear, apricot, coffee, olives and mangos (Monselise and Goldschmidt, 1982). This phenomenon occurs often in olive trees. This creates many economical and technical problems to the olive growers. Even if one hope for a good yield one year and less yield the following year, it has been demonstrated that the total yield will be anyway less in comparison with constant year production. Nutritional problems, lack of water, pest and diseases, frost in spring season are some of the possible factors favoring the alternate bearing. In this phenomena there is an overlapping of the biological fruit cycle of one year to the next year. The year of no fruits on the tree due to unfavourable conditions, there is a nice vegetative growth, this vegetative growth allows many buds to differentiate in flower buds and consequently there will be a good crop. But, this heavy crop does not allow the formation of well developed branches and in the following year the crop will be again scanty.

References

Bernier, G.1988. The control of floral evocation and morphogeneisi. *Annu. Rev. Plant Physil. Plant Mol. Biol.* **39**: 175-219.

Fabbri, A., Alerc, L. 1999. Reproductive and vegetative bud differentiation in *Olea europea* L. *J Hort Sci Biotechnol* **74**(4): 522-527.

Farinelli, D., Boco, M., Tombesi, A. 2006. Results of four years of observations on self – sterility behaviour of several olive cultivars and significance of cross – pollination. Proceedings of Second International Seminar on "Biotechnology and quality of olive tree products around the Mediterranean Basin – Olivebioteq 2006", 5 – 10 November 2006, Marsala – Mazara del Vallo Italy, Vol. I: 275-282 Alcamo.

Fernandez-Escobar, R., Benlloch, M., Navarro, C., Martin, G.C. 1992. The time of floral induction in the olive. *J Amer Soc Hort Sci* **117**: 304-307.

Goldschmidt, E.E. 2005. Regolazione dell'alternanza di produz-ione negli alberi da frutto. *Italus Hortus* **12**(1): 11-17.

Hartmann, H.T. 1951. Time of floral differentiation of the olive in California. *Botanical Gazette* **112**: 323-327.

Koubouris, C.G., Metzidakis, I., Vasilakakis, M.2009. Influence of cross pollination on the development of parthenocarpic olive (*Olea europea*) fruits (shotberries). *Expl Agric*. **46** : 67-76.

Monselise, S., Goldschmidt, E.E. 1982. Alternate bearing in fruit trees. *Horticultural Review* **4**: 128–173.

Nilsson, S. 1988. A survey of the pollen morphology of Olea with particular reference to *Olea europaea* sens. lat. *Kew Bull*. **43**: 303-315.

Ollivier, D., Artaud, J., Pinatel, C., Durbec, J.P. and Guérère, M. 2006. Differentiation of French virgin olive oil RDOs by sensory characteristics, fatty acid and triacyl glycerol compositions and chemometrics. *Food Chemistry* **97** : 382-393.

6
Utility

Due to rising awareness about the beneficial effects of optimal nutrition and functional foods among today's health conscious cosmopolitan societies, the worldwide consumption of olives and olive products has increased significantly (Ghanbari *et al.*, 2012), especially in high-income countries such as the United States, Europe, Japan, Canada and Australia, resulting the rapid development of olive-based products (Ryan and Robards, 1998; Vinha *et al.*, 2005). The traditional "Mediterranean diet", in which olive oil is the main dietary fat, is considered to be one of the healthiest because of its strong association with the reduced incidence of cardiovascular diseases and certain cancers (Knoops *et al.*, 2004; Trichopoulou *et al.*, 2003). The olive is a healthy oil because it is mainly characterized by the presence of high content of monounsaturated fatty acid (MUFAs) and functional bioactives including tocopherols, carotenoids, phospholipids and phenolics, with multiple biological activities (Covas *et al.*, 2006; Covas *et al.*, 2008). Such components also contribute to the unique flavour and taste of olive oil. As with other crops, the composition of olive and olive oil components varies in relation to various factors, namely cultivar, ripeness and harvesting regime, agroclimatic conditions as well as the processing techniques employed (Covas *et al.*, 2006)

The olive in general contains about 15-20 per cent in olive oil, 30-60 per cent in water and the remaining is fiber, sugars, and proteins. The pulp contains 96-98 per cent of the oil, while the seed contains only 2-4 per cent of the oil. The average fatty acid composition of olive oil is 78-83 per cent in mono-unsaturated oleic acid, 6-9 per cent in essential polyunsaturated linoleic acid, 8-15 per cent in saturated palmitic acid and 1.5-3 per cent in stearic acid. Olive oil also contains 19 mg of Vitamin E per 100 g of oil. The natural presence of Vitamin E which plays the role of nutrient and natural anti-oxidant, the important presence of monounsaturated fats and the right bearing of the essential polyunsaturated ones, make olive oil a product to be preferred and chosen. Olive oil is not a significant source of dietary fiber, sugar, Vitamin A and C, Calcium and Iron. Olive oil does not contain sodium.

Fatty Acids Composition of Olive Oil

Olive oil is rich in oleic acid, and has very less proportions of other acids such as palmitic, stearic or linolenic acid. Olive oil composition may vary according to the cultivars, genotypes, variety, altitude, weather conditions and stage of harvesting. Depending on their chemical composition, fatty acids are generally classified as saturated, monounsaturated and polyunsaturated. The percentages contained in olive oil typically range around 5 per cent, 80 per cent and 15 per cent respectively.

Olive Oil Composition and Triglycerides

The fatty acids of olive oil are esterified with glycerol to form the so-called triglycerides which are formed by combining glycerol with three molecules of fatty acid. More than different triglycerides could be formed in total, though only a few are actually found in the olive oil. The triglycerides found in greater proportion are those which contain three molecules of oleic (40-60 per cent), two of oleic and palmitic (12-20 per cent) or a linoleic (12.5-12.0 per cent) or one of stearic (37 per cent) and finally a triglyceride molecule consisting of palmitic, oleic and linoleic acid (5.5-7.0 per cent). Generally, the unsaturated fatty acids oleic and linoleic show a preference for occupying the position 2 of the triglyceride.

Unsaponifiable and Hydrocarbons

These constituents make up approximately 30-50 per cent of the unsaponifiables. The most important hydrocarbon is squalene ($C_{30}H_{50}$), which can represent up to 90 per cent of these compounds, reaching concentrations of about 1250-7500 mg/kg. These are followed by carotenoids, β-carotene being the precursor of vitamin A, the most abundant (0.9 to 5.0 mg/kg). Also present are lycopene and other linear or branched hydrocarbons, some of which have been found in the volatile fraction (phenanthrene, anthracene, pyrene) at trace levels.

Non-glyceride Esters

These include waxes, which are esters of aliphatic fatty acids with high number of carbon atoms. The maximum content is set at 250 mg/kg for olive oil and 350 mg/kg for the rest, in order to prevent fraud. However, they are not bad from a nutritional point of view. Thesterol esters are also important compounds within this group, mainly sitosterol, campesterol, stigmasterol, and other alcohols.

Aliphatic Alcohols

In aliphatic compounds, carbon atoms from C_{18} and C_{30} are joined together in straight chains. They are found between 100 and 200 mg/kg.

Alcohols and Triterpene Alcohols

Alcohols are found in proportions from 1000 to 3000 mg/kg and the triterpenes are made basically of uvaol and erythrodiol, which are used to detect mixtures, since they are more abundant in oils extracted by solvent.

Sterols

The concentration of sterols in the oil is about 2600 mg/kg. They have a characteristic composition and may be used to classify olive oils from different regions. From a nutritional point of view, the most significant is the β-sitosterol which helps to reduce the absorption of cholesterol in mammals.

Tocopherols

It is to be known as one of the most important lipid soluable natural antioxidants, and has a properties to prevent lipid peroxidation by scavenging radicals in membranes and lipoprotein particles (Esterbauer *et al.*, 1991). Four different types of tocopherol, namely α-, β-, γ- and δ-tocopherol have been reported in olive oil. The amount of main component, α-tocopherol, varies from a few ppm up to 300 ppm (Blekas *et al.*, 2002). The tocopherols have antioxidant properties protecting oils from oxidation as well as biological activity through vitamin E. They are thermolabile and disappear during the refining process.

Tocotrienols

Its chemical composition is similar to tocopherols, with one more double bond than the tocopherols in the lateral hydrocarbon chain. Olive oil contains a very small amount of tocotrienols.

Pigments

Olive oil, like other vegetable oils, contains considerable amount of pigments such as chlorophylls and carotenoids. Chlorophylls are encountered as pheophytin. Pheophytin α concentration in olive oil ranges from 3.3 to 40 ppm, while pheophytin *b* and chlorophyll *b* are present in trace amounts and chlorophyll *a* has not been detected (Psomiadou *et al.*, 1998). The main carotenoids present in olive oil are β-carotene (0.3–4.4 ppm) and lutein (trace-1.4 ppm) (Lanzon *et al.*, 1994) for instance, the concentration of carotenoids in Spanish olive oils is 3.1–9.2 mg/kg (Gandul *et al.*, 1996). Triterpene hydrocarbon carotenoids are responsible for the yellowing of oils and green table olives. The highest concentration is found in β-carotene and xanthophyll. The concentration can vary between 5 and 10 mg/kg.

Polyphenols

These compounds are widely distributed in the plant kingdom as well as in different plant parts (leaves, fruits, *etc.*). They act as antioxidants and have very interesting biological properties. Of all oils, olive virgin oil and to a lesser extent, the olive fruit and olive pomace contain it. It is virtually lost during refining, mainly during the neutralization and bleaching stages. Both lipophilic and hydrophilic phenolics are distributed in olive fruit. The main lipophilic phenols are cresols while the major hydrophilic phenols include phenolic acids, phenolic alchohols, flavonoids and secoiridoids; they are present in almost all parts of the plant but their nature and concentration varies greatly between the tissues (Boskou 1996; Covas *et al.*, 2006).

Other Important Chemical Substances which Present in Various Chemical Forms

Alcohols, phenolic compounds (hydroxytyrosol and tyrosol), free phenolic acids (vanillic, p-coumaric, caffeic, *etc.*) esterified derivatives of caffeic acid (verbascoside) or acid elenolic (oleuropein, oleuropein aglycon, and so on.), flavonoids (luteolin, rutin, quercintina, *etc.*). It appears that oleuropein is a specific compound in olives responsible for the bitter taste and the inhibition of microorganisms. The phenols are relatively simple compounds produced from degradation by different mechanisms (enzymatic or chemical) of the original compounds present in the fruit. There is a great diversity of them, and even though the molecule is relatively simple they have similar antioxidant and biological activities. Currently, most of the olive oil nutritional properties are attributed to the unsaponifiable fraction, and some of them more specifically to the polyphenols. Content can range from 50 to 1000 mg/kg. Olive oil is liquid at room temperature. It has some vegetable oil and olive fruit characteristics.

Nutrition Facts

Fat Content

Fat is an important source of energy for the human body. It supplies 9 kcal per gram of fat consumed. However, as olive oil is not easily absorbed when frying the energy value is different. Olive oil has an unsaponifiable fraction that should not be considered when estimating its energy value. In any case, the triglycerides are intended to supply this energy by using their metabolism of glycerol and fatty acids itself. Part of the ingested fatty acids becomes part of the fat deposits that make up the body energy reserves. Its composition depends on the type of fat eaten, so there is a similarity between ingested and stored. Therefore, if a person usually eats olive oil (rich in oleic acid), in the inter- digestive periods or when the body requires the mobilization of the adipose tissue, she/he will be using a fat which is predominantly oleic acid.

Essential Fatty Acids Supply

Essential fatty acids are Linoleic and α-Linolenic. Olive oil contains 8-9 per cent and 1 per cent respectively. These amounts are enough to meet the nutritional requirements of the organism. On the other hand, the ratio between them in the oil is fairly balanced, for what is currently considered a healthy diet in relation with ω-9/ω-3.

Effects on the Digestive System

In olive oil, especially oleic acid has regulatory effect on the digestive system by stimulating different hormones or peptides. Thus, olive oil compared to other oils is clearly differentiated in the amount of oleic acid contained.

Effects on the Composition of the Cell Membrane

Because of olive oil posses high content in oleic acid, its intake leads to less oxidized LDL lipoproteins and, therefore, less atherogenic. The fatty acids in the phospholipids that form the cell membranes also have a lower unsaturation degree. This unique composition gives olive oil its special chemical and physical characteristics of fluidity and permeability, which affect multiple vital functions such as respiration, oxidative phosphorylation, signal transduction receptor inside the cell, passage of metabolites, *etc.*

Modulation of Eicosanoid Synthesis

Oleic acid intake modulates the biosynthesis of eicosanoids, involved in many functions such as smooth muscle contraction, platelet aggregation, diverse inflammatory causes, and so on.

Nutritional Value of Minor Components

Olive oil is a major source of vitamin E and plays a decisive role defending the cells due to its antioxidant properties. A daily intake of 50 g of olive oil covers 100 per cent of the recommended dietary intake of this vitamin needed for men and women. But vitamin E is partially lost in the refining process, thus the contribution of this vitamin in olive oil blends (virgin and refined) is significantly lower and will depend on the proportions used in their preparation. Extra virgin olive oil as well as in virgin oil, (although to a lesser extent) are rich in vitamin A, and more specifically the provitamin α-carotene and contribute as a daily source of the this vitamin. Beside theses vitamins the phenolic substances also exist in olive oil and act mainly as antioxidants and its action and their main effects are related to the cardiovascular system. In vitamin E, they are present at the highest possible proportion in the extra virgin or virgin olive oil and in lower proportion in olive or olive pomace, depending on the proportion of virgin olive oil used in the mixture of oils prepared.

The important components of olive oil are triglycerides and fatty acids which are the saponifiable fraction, and a number of compounds (0.5-2.0 per cent) that constitute the so-called unsaponifiables are very useful for the oil stability and flavour. One of its main properties of olive oil is its high content of oleic acid (75 per cent average). The properties of olive oil depend on the variety of olives used, the way in which the oil was processed and the storage procedures.

Properties of Olive Oil, Organoleptic

The organoleptic qualities of olive oil (*i.e.* flavour and odour) are generally evaluated as per specific standard procedure established by the International Olive Oil Council. The main aim of the testing is to determine the positive and negative qualities of the oil which ultimately related to olive oil health benefits.

The main positive qualities are fruitiness, bitterness and spiciness.

☆ **Fruitiness**

It is mainly contributing to the fresh fruit taste of olive oil which indicates the healthy, fresh olives that may be ripe or unripe. This is the most important quality for the organoleptic evaluation and, if absent, the olive oil cannot be classified as "extra virgin or fine virgin" olive oil.

☆ **Spiceness**

It majorly imparts the piquant sensation in the throat; this is usually observed in "agoureleo" (made from green olives) and it is due to the action of phenic substances on the tip of the trigeminal nerve. This sensation disappears a few seconds after tasting.

☆ **Bitterness**

It is a characteristic flavour due to the use of green or greenish olives and, depending on its intensity, may be pleasant or unpleasant – but in no case is it considered to be a negative property.

Negative qualities include "atrojado" (which appears in Spanish olive oils), mould, sediments, winey, metallic, rancid, burnt, hay-wood, fat, brine or other flavours. The virgin olive oil colour ranges from gold to dark green, depending on the variety of olives used, but does not mean that its quality is better or worse. The positive attributes must be well balanced and all flavours properly blended. Olive oil (not extra virgin or virgin) is not subject to tasting panels as after refining it has minimum organoleptic properties compared to any virgin or extra virgin olive oil. The taste or smell in olive oil which is not virgin, comes from the small amounts of olive oil or extra virgin contained in it. Properties can vary depending on the olive type (IOC, 2015)

Health Benefits

Olives are typical in their fat quality, because they provide almost three-quarters of their fat as oleic acid, a monounsaturated fatty acid. The high monounsaturated fat content of olives beneficial in heart related ailments by decrease in blood cholesterol, LDL cholesterol, and LDL:HDL ratio. All of these changes lower risk of heart disease.

Anti-cholesterol Properties

The monounsaturated fats and polyphenols helps in reduction in oxidation of cholesterol which leads to powerful protective and preventive effect against atherosclerosis and related cardiovascular diseases, such as stroke and heart attack.

Antioxidant and Anticancer Activities

The antioxidants like polyphenols, vitamin E and beta-carotene are the most beneficial substances present in olives. The polyphenols are good and act as agent against the free radicals, prevent cancer, premature aging, heart disease and other degenerative and chronic conditions.

Bone Health

Olives are rich in vitamin D, calcium and phosphorous, which all play a critical role in bone growth, remodelling and maintenance and help prevent bone conditions such as rickets in children and osteoporosis in adults.

Heart Health

The olives and oil has impact in preventing blood clots formation and promoting vasodilatation. This results in decreased heart work and improved heart function.

Purifying Effect

It is healthy for liver's and the intestinal functions since its having high content of fibber which help cleanse the colon, as well as prevent or fight constipation resulted detoxification and excretion of toxins from the body, with consequent improved function and health of the whole body.

Restorative Properties

The olive oil has high content of beneficial minerals and act as natural alternatives to multi-mineral supplements used to give the body more energy, strength and nutrients.

Skin Health

It helps in preventing the damage caused by free radicals on skin tissue. Beside these olives also contain relatively high amounts of beta-carotene, the precursor of vitamin A, which play an important role in stimulating skin regeneration and providing skin protection.

Vision

Beside skin health, but is also essential for normal vision, especially in low light, as well as for eye health and integrity.

Anti-Inflammatory Benefits

The studies showed that whole olives extract has properties to act as anti-histamines at a cellular level by blocking special histamine receptors (called H1 receptors) which provides also anti-inflammatory benefits. In addition to their antihistamine properties, it also helps in lowering the risk of unwanted inflammation by lowering levels of leukotriene B4 (LTB4), a very common pro-inflammatory messaging molecule. Oleuropein present in olives which have the potential to decrease the activity of inducible nitric oxide synthase (iNOS). iNOS is an enzyme whose over activity has been associated with unwanted inflammation.

References

Blekas, G., Psomiadou, E., Tsimidou, M. 2002. On the importance of total polar phenols to monitor the stability of Greek virgin olive oil. *Eur. J. Lipid Sci. Technol.* **104**: 340–346.

Boskou, D. 1996. History and characteristics of the olive tree. In: Boskou D., editor. Olive Oil Chemistry and Technology. Am. Oil Chem. Soc. Press; Champaign, IL, USA.

Covas, M.I. 2008. Bioactive effects of olive oil phenolic compounds in humans: Reduction of heart disease factors and oxidative damage. *Inflammopharmacology* **16**: 216–218.

Covas, M.I., Nyyssonen, K., Poulsen, H.E. 2006. The effect of polyphenols in olive oil on heart disease risk factors. *Ann. Int. Med.* **145**: 333–343.

Esterbauer, H., Dieber-Rotheneder, M., Striegl, G., Waeg, G. 1991. Role of vitamin E in preventing the oxidation of low-density lipoprotein. *Am. J. Clin. Nutr.* **53**: 314S–321S.

Gandul-Rojas, B., Minguez-Mosquera, M.I. 1996. Chlorophyll and carotenoid composition in virgin olive oils from various Spanish olive varieties. *J. Sci. Food Agric.* **172**: 31–39.

Ghanbari, R., Anwar, F., Khalid, M. A., Gilani, A. and Saari, N. 2012. Valuable nutrients and functional bioactives in different parts of olive (*Olea europaea* L.)—A Review. *Int J Mol Sci.* **13**(3): 3291–3340.

IOC. (2015). http: //www.internationaloliveoil.org/

Knoops, K.T., de Groot, L.C., Kromhout, D. 2004. Mediterranean diet, lifestyle factors, and 10-year mortality in elderly European men and women. *J. Am. Med. Assoc.* **292**: 1433–1439.

Lanzón, A., Albi, T., Cert A. 1994. The hydrocarbon fraction of virgin olive oil and changes resulting from refining. *J. Am. Oil Chem. Soc.* **3**: 285–291.

Psomiadou, P., Tsimidou, M. 1998. Simultaneous HPLC determination of tocopherols, carotenoids, and chlorophylls for monitoring their effect on virgin olive oil oxidation. *J. Agric. Food Chem.***46**: 5132–513.

Ryan, D., Robards, K. 1998. Phenolic compounds in olives. *Analyst* **123**: 31R–44R.

Trichopoulou, A., Costacou, T., Bamia C., Trichopoulos D. 2003. Adherence to a Mediterranean diet and survival in a Greek population. *N. Engl. J. Med.* **348**: 2599–2608.

Vinha, A.F., Ferreres, F., Silva, B.M., Valentão, P., Gonçalves, A., Pereira, J.A., Oliveira, M.B., Seabra, R.M., Andrade, P.B. 2005. Phenolic profiles of Portuguese olive fruits (*Olea europaea* L.): Influences of cultivar and geographical origin. *Food Chem.* **89**: 561–568.

7

Olive Improvement

The genetic improvement programmes of olive are directed to solve pomological and commercial problems such as:

- ☆ The production of self-fertile plants.
- ☆ The regulation of fruit ripening and increase of oil content and quality.
- ☆ The production of plants with parthenocarpic fruits.
- ☆ Increase of cold and salt tolerance.
- ☆ Modification of the vegetative habits.
- ☆ Resistance to biotic stress.

1. Constraints in Olive Breeding

a. **Poor Fruit Setting is a Major Problem:** Major obstacle is the low fruit set of most olive cultivars. In arid area with frequent periods of moisture stress during late winters or spring, intermediary flowers containing rudimentary but non-functional pistils may develop. Such flowers are incapable of bearing fruits. The ability of perfect flowers to bear fruits is further affected by adequate pollination and post pollination abscission. The extent of pistil abortion is cultivar dependent and prevailing weather conditions. Each olive cultivar has been found to show a specific extent of pistil abortion even under normal or ideal weather conditions. Thus the environmental factors, genetic constitution and cultural practices seem to have a pronounced effect on the expression of this phenomenon. Generally three types of ovary aborted flowers are found in olive namely:

   - ☆ Type –I : Indicate lack or absence of ovules.
   - ☆ Type –II : Indicate under developed style and outer tissue of the ovary and ovules.

- Type-III : Indicate completely underdeveloped female organs.
- The fruit set is very low in olive. Olive produces redundant flowers relative to its yield potential and fruit set is very low. The ovary abortion does not affect pollen production in olive. In this species, large-fruited cultivars have greater ovary abortion, but similar number of flowers, leaving the male function probably unaltered.

b. **Low Seed Germination, and Slow Initial Seedling Growth Rate.**

c. **Presence of Self-Incompatibility Creates a Problem in Breeding:** Others hindrances in hybridization programmes being the high degree of self-incompatibility of most cultivars disfavoring selfing, ultimately resulting in decreased opportunities for developing homozygous progenies, and makes the paternity of progenies uncertain to some degree. Self-incompatibility in olive has been reported by many workers. Self-incompatibility inhibits or delays the process of pollen tube growth leading to lower amounts of fertilization. Cultivars differ for the levels of self-incompatibility, which can be either self-compatible to completely self-infertile ones. Prevailing climatic and environmental conditions before and during the bloom and fruit set periods affect the degree of self-compatibility and yield. Temperature of 30°–35°C can have inhibitory effects on the germination and growth of the pollen tube and leading to increase in the level of self-incompatibility, cross-incompatibility in some cross combinations and in general, pollen incompatibility. Pollen tube growth under self-fertilization is generally slower than under cross-fertilization. The ovule longevity becomes critical due to slower tube elongation under self-pollination as degeneration of embryo sac begins before the pollen tube actually reaches it.

d. Factors like high level of heterozygosity, masks the expression of recessive genes; scarce knowledge about the inheritance characters and their hereditability make the cross breeding complex.

e. Cross breeding programmes in olive have to face some practical problems which involve the aspects of olive biology and morphology: flower morphology (small flowers, very small and closed anthers, delicate pistils) which makes the emasculation and controlled cross-pollination very difficult.

f. **Long juvenile phase**: Most of the fruit crops have a long juvenile phase; likewise olive is no exception that starts bearing in 10-20 years, depending on the progeny. Progenies of Coratina, Picholine, Tanche and Manzanilla have long juvenile period whereas, in the progenies of Verdale, Gordal and Leccino bud differentiate into flowers is induced very early, influence of mother plant and pollinizer on juvenility of progeny has been advocated by some workers.

   Selection process in crops like olive is complicated because productive and qualitative traits (oil or table olives) do not possess spatial and temporal stability.

2. Long-term Breeding Objective

a. Regulation of Fruit Ripening and Increase of the Oil Content and Quality

A lot of cultivars have good agronomic characteristics, but they present high water content of the fruits, early dropping, early ripening and. This problem can be overcome by generating transgenic plants with a reduced ethylene biosynthesis, by using an antisense gene. At present, two molecular strategies can be used to modify oil composition and content.

- ☆ Alteration of the major fatty acid level by suppressing or expressing a specific key enzyme in lipid biosynthesis.
- ☆ Creation of an unusual fatty acid.

The oleic acid (C18:1) can be enhanced by more than threefold (from 24 per cent to 80 per cent) by anti-sense suppression or co-suppression of oleate desaturase. This kind of technologies have been adopted to increase stearate (C18:0) by up to 30 per cent both in canola and soybean oils. Unusual fatty acids can be produced in one plant by transferring a gene encoding the specific biosynthetic enzyme. An example can be seen in canola which naturally does not produce laurate (C12:0), while a new transgenic genotype does contain laurate. The oil content of some nut crops used for cosmetics, such as almond, could be increased or their composition could be modified by these techniques.

b. Parthenocarpy

Alternate bearing habit is a major limiting factor for olive production; this may be due to lack of pollination in areas with less intensive olive cultivation. Blossom synchronization between cultivars also varied over the season. This problem is more persistent in the regions where the plantation intensity is low and the climatic conditions are unfavourable. A parthenocarpic trait in olive is an important trait which may allow better fruit development and subsequently boost the yield. In some cases it has been observed the tendency in olive to have natural parthenocarpy, under adverse environmental conditions but the fruits remain very small. The introduction of parthenocarpy Arabidopsis gene, already successful in several herbaceous species, including aubergine (Rotino *et al.*, 1997) may overcome the pollination problem.

c. Cold and Winter Chilling Tolerance

For commercial cultivation cold tolerant genotypes are more desirable for the cold prone areas. In other species, transgenic plants with gene encoding the antifreeze protein have been generated and the over expression of SOD (Super Oxide Dismutase) gene has also been demonstrated to repair frost damaged cells. Over expression of *Arabidopsis CBF1* gene enhances freezing tolerance by inducing genes associated with cold acclimatisation and could also be useful in olive.

d. Alteration of the Vegetative and Reproductive Habits

The olive breeding is aimed to impart dwarfness and changes in canopy architecture for developing dwarf and semi-dwarf genotypes with shorter and numerous shoots, that's more suits to high-density orcharding. In addition developing genotypes with an extensive root system and/or and reduced water consumption could be useful for drought condition Some genes of the TL-DNA of *Agrobacterium rhizogenes* that affect plant growth and development have already been introduced in olive, kiwi, cherry and peach. In all these species, the reduction of plant size and apical dominance is confirmed and the current field trials will show their agronomic value. Beside this, the reduction of flower numbers observed in kiwi (Rugini *et al.,* 1997) may contribute to solving the waste of energy and to reducing the alternate bearing. The modern biotechnological tools plays an important role in modifying the growth and reproductive behaviour and have an ability to modify plant receptors in order to change the light perception. The transformation with phytochrome genes (*phyA*, and/or *phyB*, sense or antisense), which, together with other photoreceptors, control plant development (circadian rhythms, apical dominance, blossoms, growth and fruit ripening, photosynthesis products distribution, development of photosynthetic systems, transpiration control and hormone synthesis) may contribute to developing plants with high agronomic value and suitable for very high density planting.

e. Pests and Diseases Resistance

The challenge of using molecular biology in this area is to generate broad resistance mechanisms that have been difficult to achieve with classical plant breeding approaches. For enhancing protection from fungi, in addition to osmotin gene or an association of other genes, as already introduced in olive, other genes may be introduced after first testing the gene product against the main fungi of olive. *Stilbene syntase* gene, ribosomeinactivation protein gene, *glucose oxidase* gene and genes encoding hydrolytic enzymes such as chitinase and glucanase, which degrade fungal cell wall component, may contribute to protecting olive. Particular attention is given to Polygalacturonase-Inhibiting Proteins (PGIPs) that specifically inhibit the activity of endo-polygalacturonases released by fungi on invasion of the plant cell wall. Genotypes with less susceptibility to *Pseudomonas syringae* pv. *savastanoi* should be tried. Thionin and its synthetic analogue MB39, attacin E and cecropine seem to enhance resistance to bacteria in other species. In the battle against insect attack, the *Bt* gene from *Bacillus thuringensis* has successfully been introduced in other species with encouraging results, and could be an alternative way to prevent olive fly damage.

Immediate Breeding Targets

a. The olive improvement mainly focuses on increasing the yield, resistance to biotic factors *i.e.* diseases and insect pests, wider adaptability under specific cultural practices and environmental conditions. Yield maximization in olive is a function of factors like, self-fertility, abundant flower differentiation and fruit set, reduced fruit drop, low alternate bearing and short unproductive period.
b. For olives intended for table purpose, large fruits, high flesh percentage, preferred organoleptic characteristics, and suitability to processing (low oil content, high quantity of reducing sugars, good pulp firmness, high

skin resistance, with stable and uniform colour) are the major traits of concern. Besides these, smooth stone and easy separation from the pulp from pit also form the objectives of any breeding programme.

c. The quality of olive oil is determined by its rich aroma, storage behaviour, antioxidants like tocopherols and polyphenols, saturated and unsaturated acids ratio *etc.* So, development of cultivars with above parameters may be an objective of crop improvement.

d. Breeding for resistance to major pests (*Bactrocera oleae, Prays oleae* and *Saissetia oleae*), fungal diseases (*Verticillium dahliae, Spilocaea oleaginea*) and bacterial diseases (*Pseudomonas syringae* pv. *savastanoi*) and tolerance to cold temperatures, drought and salinity also hold tremendous potential.

e. A cultivar is considered desirable if it possesses easily detachable larger fruits of high quality. Compact and erect tree habit, strong fruiting shoots, are important to facilitate pruning and harvesting.

f. Olive breeding with the objective to develop plants that invest less in male fitness *i.e.* reduced flowering, will save plant energy, diverting it to produce more fruits. This appears reasonable since the energetic cost of flowering is not negligible in olive.

g. High rooting ability is essential for propagation purposes, both for cultivars and for rootstocks. The main issue in olive breeding is its adaptability to wider climatic and soil regimes.

3. Cytogenetics

Olive (*Olea europaea* L.) is believed to be indigenous to the Mediterranean region where it grows abundantly as thick forests and is endemic to the region. The olive (*Olea europaea* L.) is a pre-historic species that represents one of the most important fruit trees in the Mediterranean basin. The monophyletic olive family, *Oleaceae* is a vast family of 30 genera and 600 species *Olea europaea* L. is the main cultivated species of the family. The *Olea* genus comprises 30 species and has spread over Europe, Asia, Oceania and Africa. The Mediterranean olive, *Olea europaea*, subspecies *europaea*, which includes wild (*Olea europaea* subsp. *europaea* var. *sylvestris*) and cultivated olives (*Olea europaea* subsp. *europaea* var. *europaea*), is diploid species (2n = 2x = 46). Cultivated olives are nearly all diploids with 2n=2x=46, occasional triploids and tetraploids have also been reported.

4. Inheritance Pattern/Linkage of Characters

The heritability studies in olive are based on the germplasm characteristic rather than the specific tests involving progeny evaluations. The large genome size and highly heterozygous nature of the genotypes make the cytogenetic studies very complex leading to a limited knowledge about the parental value and heritability of traits. The information derived from different hybridization programmes is still limited and not always coherent which reduces the efficiency of olive breeding programmes, with genotypes/seedlings selection ratio of about 0.4 per cent. The correlations based mainly on the traits observed in olive germplasm are given in Table 4.

Table 4: The Inheritance/Correlation among Major Horticultural Traits in Olive

Trait	*Notes*
Tree vigour	**Correlated to:** Abscisic acid content in root, branches and leaf, Number of cells in the palisade per unit of leaf area, Stomatal density per unit of leaf area, Cultivar can be evaluated for vigour only when trees are at least 6 years old.
Early bearing	**Correlated to:** Pale leaf colour, Productivity Picholine, Leccino, Ascolana Tenera, Pendolino, Coratina and the new cultivars I-77 and Fs-17 are known for their early bearing. In Portugal, early bearing cultivars are: Azeitoneira, Negrinha, Galega Vulgar, Cordovil de Serpa, Picual and Branquita.Verdale, Gordaland, Leccino and Picholine seedlings have shorter juvenility.Coratina, Picholine, Tancheand Manzanilla seedlings have long juvenility. Both the parents and the type of crossing affect earliness of blooming in seedlings; a correlation between short juvenility of seedlings and early bearing of parents was hypothesized. Other authors state that the length of juvenility is affected mainly by the maternal parent.
Productivity (oil)	**Correlated to:** Cultivar and growing environment, Pulp to pit ratio and drought sensitivity of the cultivar, Leaf area index (LAI).
Germinability	**Correlated to:** Maternal parent characteristics. Maternal cultivars in decreasing order of ability to transmit the character are Picholine > Leccino > Manzanilla > Verdale > Tanche. Paternal cultivars in decreasing order are Leccino> Frantoio> Picholine> AglandauCultivars Manzanilla, Kalamon and Chalkidikis confer good germinability to their seedlings.
Rooting ability	This character is highly variable in olive, with results varying from zero to 100 per cent of rooting of cuttings, depending on the cultivar. Good rooting was obtained with Ciliegino, Pesciatino, Razzaio, Rossellino, Manzanillo, Weteken, Bashika I, Ashrasiand Labeeb II. Comparing seedlings from 127 crossing combinations, a greater root development was obtained from crossings involving Picholine.
Fruit: size, weight, skin colour	Fruit size is affected by environment and fruit load. The cultivar Grossanne gives high and uniform fuit size to its progeny The crossing Picholinex Grossannehas a high percentage of seedlings with fruits > 3 g Green colour is more often obtained in the Grossanne x Frantoio progeny, reddish colour in the Picholine x Coratina progeny, purple in the Picholinex Grossanne, and black, which is quite less frequent, in the Leccinox Aglandau and Leccinox Gordal Early fruit coloration is obtained in the progenies of Coratinax Picholine, Picholinex Manzanillaand Bouteillanx Coratina.
Fruit: flesh firmness	**Correlated to:** Histological structure of the cuticle, epidermis and ipodermis. Highest firmness is obtained with cultivar Palma and lowest with Corsicana and Gordal.

Contd...

Table 4–*Contd...*

Trait	*Notes*
Oil: acidic composition	Mainly under genetic control.
Resistance to *Spilocaea oleaginea*	**Correlated to:** Thickness of palisade in leaves. Large differences exist even within the same cultivar, as was found in a population of Tondo Sassarese, a highly susceptible cultivar, but with interesting sources of resistance.
Resistance to *Bactrocera oleae*	**Correlated to:** Early ripening, Size and shape, water and oil content of fruit.
Resistance to frost	**Correlated to:** Plant physiological mechanisms, Leaf water potential High sugar content (saccharose and malthose) and phosphorus, greater perossidase and polyphenolossidase activity, greater acidity in the cell lymph, and lower redox potential, Stomatal density, Stomatal perimeter, Sugar/nitrogen ratio, Photosynthetic capacity, Phenolic compound release and browning of leaf tissues, Electric conductivity and Soluble protein content in leaves. Cultivars Leccino, Nostrale di Rigali, Laurina, Moresca, Negrera, Ogliarola, Itrana, Tanche, S. Caterina and the French cultivar Bouteillan are considered particularly resistant while cultivars Frantoio and Moraiolo are considered sensitive Great variability exists within the cultivars Leccino and Moraiolo depending on the different clones.
Resistance to drought	**Correlated to:** Greater root development and ability of the plant to adjust its water potential relative to the water potential of the soil.
Resistance to salinity	Very variable trait, depending on cultivars. Cultivars that appear more tolerant are: Frantoio, Arbequina, Nevadillo, Lechin de Sevilla, Jabaluna, Cañivano and Picual, Pocama, Hamed, Verdale, Picual, Megaritiki and Amphissis, Kothreiki and Kalamata.

References

Rotino, G. L., Perri, E., Zottini, M., Sommer, H., and Spena, A. 1997. Genetic engineering of parthenocarpic plants. *Nature biotech* **15**: 1398-1401.

Rugini, E., Caricato, G., Muganu, M., Taratufolo, C., Camili, M., and Cammilli, C. 1997. Genetic stability and agronomic evaluation of sixyear- old transgenic kiwi plants for *rolABC* and *rolB* genes. *Acta Hort* **447**: 609-610.

8

Biotechnological Advances in Improvement

The olive has a great importance in the international food market due to the health benefits and culinary qualities of its products. In recent times its cultivation is gaining popularity in India too. Beside this, ever increasing scientific evidence of the beneficial properties of the regular consumption of olive oil has encouraged an increase in its consumption worldwide.

For most of the crop species, including the olive, advances in field performance depend on knowledge obtained by modern biotechnology and molecular biology, which supports new and more accurate approaches in many research fields, including genetics, plant pathology, physiology and biochemistry, entomology and plant nutrition. Advanced biotechnological tools have applications in diverse fields such as foods, clothing, cosmetics, pharmaceuticals and a number of industrial materials and became an essential tools for increasing the quality and productivity of fruit crops/species in worldwide due to achieve significant gains. The technologies developed in the fields of biotechnology and molecular biology can be combined with traditional approaches, such as conventional breeding, for the potential development of new genotypes with superior traits more rapidly and aggressively.

The Different Biotechnological Approaches

In vitro Culture of Olive

The multiplication of the olive in *in vitro* condition remains a major focus is due to its huge advantages as a large-scale cloning technique compared to conventional methods such as rooting of semi hard wood stem cuttings and grafting. The olive tree showed a high level of cross pollination and out crossing therefore the production of seedlings from seeds is not practiced for commercial use. The seed propagation mainly used with the aim of genetic breeding. The plants obtained from seeds produce genotypes different from that of the mother plant, and consequently, they may not exhibit the same desirable qualities as the mother plant due to the genetic segregation. In addition, the occurrence of a lengthy juvenile phase due to seed propagation delays the production of flowering, fruiting and the first harvest. The stem cuttings and grafting is an asexual method of propagation are for the production of olive plantlets, although these methods has low efficiency and are influenced by environmental factors such as genetic constitution, nutrition and health status of plant material utilized. Given these limitations, the *in vitro* clone propagation of olive may assume an important role by reducing or even overcoming some of these limitations (Rugini and Caricato, 1995).

The *in vitro* culture is also an essential tool for physiological and biochemical studies of in fruit crops, including wild relatives whose potential remains unexploited. The *in vitro* multiplication and propagation techniques has widely used for research purposes and is well documented in the olive and standard protocols also standardized but its commercial application remains challenging and almost unexplored.

Micropropagation of Olive

This technique is a now a day's most common in tissue culture and transgenic plant development with high practical impact in fruit science. Micropropagation is used frequently in difficult to root fruit crop species. Due to its practical importance of this technique and low cost have stimulated its application in horticulture.

Advantages of Micropropagation

- ☆ Uninterrupted production of quality plant material round the year.
- ☆ Free from seasonal dependence.
- ☆ Large number of plant from limited space.
- ☆ Large-scale and faster multiplication of elite genotypes from limited stock plants.
- ☆ Generation and maintenance of quality and healthy plants.

There is different method to produce plants from either zygotic or mature tissues from different cultivars based on explants such as micropropagation techniques:

1. By auxiliary buds stimulation
2. By organogenesis
3. By somatic embryogenesis.

Micro-propagation by Axillary Bud Stimulation

Protocols for *in vitro* propagation of the olive cultivars were reported many years ago (Rugini, 1984). This technique allows high quality production and rapid growth of the plants, which are pathogen free on the surface and in the vascular system. Rooting occurs in high percentage, even in cultivars considered recalcitrant's when propagated from stem cuttings like 'Nocellara etnea'. The rate of growth is superior to plants derived from stem cuttings; they produce flowers and behave in the same manner as plants derived from suckers or somatic embryos. Only micropropagated plants from mature material, from suckers or from somatic embryos (even if produced *in vitro* from adult material), spend the period of over 2 years under field conditions in respect to those derived from mature shoots. The genetic stability of the *in vitro* plants can be tested by PCR-RAPD analysis and with agronomic and morphological observations.

Micropropagation by Organogenesis

Organogenesis is the process in which the undifferentiated cells of any tissue differentiate an organ (buds or shoots, roots, flowers etc), directly from the cells of the tissue (direct organogenesis) or from the cells of a callus formed from any tissue (indirect organogenesis). The differentiated shoots will be induced to produce roots like those derived from pre-formed buds. Generally an only cell is involved in the regenerative process. This type of regeneration would facilitate the isolation of stable genotypes from chimeric tissues. The organogenesis was obtained in olive tree from zygotic tissues and from mature tissues of a cultivar, directly or through callus formation. A high frequency of organogenesis was obtained from cotyledonary fragments close to the embryos of mature seeds and sections of hypocotyls (Bao *et al.*, 1980). Nevertheless, petioles of leaves derived from *in vitro* proliferated shoots are also able to regenerate buds, even though in low frequency, not more than 40 per cent, in the function of a cultivar, of the position of the leaf on the shoot, of the substrate, as well as of the quality of the shoots during the phase of proliferation (Mencuccini and Rugini, 1993). Commonly the organogenesis cannot be considered a method of multiplication because from zygotic tissues it is not possible to reproduce the features of the mother plant due to the high heterozygosis of the species and because the efficiency of regeneration of a mature cultivar is too low. The use of this technology is essential as the first step in obtaining somatic embryogenesis.

Micropropagation by Somatic Embryogenesis

In the olive, somatic embryogenesis has been successfully by achieved from several tissues of both zygotic and maternal origin: immature and mature zygotic embryos, seedling tissues and leaf petioles of cultivars. Two to two and a half months after fertilisation, is the best period to collect fruits to extract zygotic embryos, because during this period they show the highest somatic embryogenesis potential. However, as well as the date of embryo collection, other factors such as growth regulator quality and light should be taken into consideration, because they are critical in embryo formation. The capacity of the immature zygotic embryos to form somatic embryos may be extended for at least two months by storing at 14 to 15 °C, the fruitlets collected 60-75 days after full bloom. Under these conditions, although the small embryos continue to develop, they do not, however, lose their embryogenic capacity, contrary to the corresponding embryos collected from fruits left on the plants. The first report on induction of somatic embryogenesis in the olive was achieved by wounding the roots still attached to the seedling; the callus produced in the wounded zone produced embryos which converted into plantlets (Rugini and Tarini 1986). Callus from segments of non-germinated mature embryos could produce somatic embryogenesis. Calli derived from rootlets give the highest somatic embryogenesis. The difference in the amount of embryogenesis depends on the origin of the explant (proximal, medium and distal part of the embryo) and on the last of the period of permanence of explants on callus induction medium

Production of Pathogens Free Plants

In olive propagation, almost always in asexual way, not only fungal and bacterial diseases but also viral diseases contribute to their spread. The diffusion of pathogen free olive material of commercial scale always been a challenging task for researchers and propagator work. The protocols for olive multiplication *in vitro* are available and can be applied to a wide number of olive cultivars. From the *in vitro* grown shoots it is possible to take the meristem which is well known to be virus free.

Micrografting

The olive presents difficulty to develop shoots from meristem on a common mineral substrate if it is not taken from *in vitro* grown shoots during the multiplication phase. Micrografting could represent a valid method of support to this technology in the case in which not all the genotypes are able to develop or to rejuvenate recalcitrant genotypes in the conditions *in vitro* in order to improve their adaptation to *in vitro* conditions.

Rhizogenesis (rooting) is considered as a critical process and it influences the acclimatization of the seedlings in an external environment. Seedlings with weak root systems have high mortality because they are not able to adapt their root performance to balance the uptake of water with evapotranspiration without undergoing drying. There is a relation between the internal level of auxins and cytokinin, which is responsible for initiating the rooting process. Other variables that can affect the process of adventitious root formation in olive are an excessive osmolarity of the culture medium, low availability of reserve substances in the tissue, lack of adequate nutrients, contamination by pathogens, juvenility of the materials used, thickness and lignification of the cell wall. All of these factors, individually or in combination, can completely inhibit the process of rooting.

However, the vegetative propagation of olive tree and other woody species by conventional methods has some drawbacks, mainly due to seasonal influences and the demand for large areas to produce buds for grafting and stem cuttings for plantlet rooting. The bottleneck in commercial cultivation of olive in India is the absence of sufficient certified olive plantlets with adequate phytosanitary quality and genetic identity. In addition, the lack of knowledge in strategic areas of olive cultivation and production, such as propagation, genetic breeding, irrigation and fertilizer management, disease and pest control, and post-harvest and fruit/oil processing, is also an obstacle for the cultivation of this crop. Cost is another consideration in the installation of olive micropropagation facilities. The critical step that should be considered in *in vitro* olive cultivation, whether for commercial or research purposes, is the development of adequate protocols. Similar to the rooting of stem cuttings, *in vitro* rooting is very dependent on the genetic background and is strongly influenced by the olive variety used. The influence of the genetic the intricate metabolic processes involved in plant development. Despite evergreen nature of olive trees, a significant reduction in growth development occurs during the winter season, mainly in locality with harsh winters. Olive explants exhibit variable responses to *in vitro* multiplication based on the seasons in which they are collected. During the exponential multiplication of olive cultivated *in vitro*, a satisfactory rate of genetic homogeneity must be maintained. During the multiplication phase, the occurrence of somaclonal variation must be avoided.

Somaclonal Variation

Somaclonal variation arises from genetic or epigenetic events and might be induced at a high rate by the stress during the successive subculture of explants *in vitro*. The effects of somaclonal variation occur directly on the deoxyribonucleic acid (DNA) due to the breaking of chromatin, the unbalanced migration of the chromosome during mitosis, inversions, deletions, translocations, duplications and changes in the nucleotide sequence of the DNA (mutations), which are transmitted in a hereditary manner to the progeny. Other possible reason might be due to variation are derived from changes in the methylation pattern of DNA and chromatin condensation, the formation of stable complexes with histone molecules and synthesis of non-codifying RNA (Interference RNA), which can result in the loss of gene function (gene silencing). These events are known as epigenetic effects and are also induced by environmental effects, such as the stress caused by *in vitro* culture. However, the epigenetic changes do not alter the nucleotide sequence of the DNA, and in plants, they are transferred asexually through vegetative propagation.

In order to achieve this target it is indispensable to have a very efficient method of regeneration and possibly from mature material of affirmed cultivar aimed to improve certain characteristics. Few cultivars, including cv. Canino, are able to regenerate from somatic embryos. In order to increase the genetic variation in a species, it is a common technique to grow the embryos or the embryogenic masses to ionising radiations, fungal filtered cultures, or filtered toxins of some pathogens or to other selective pressures that induce changes or allow the selection of changing material.

Changes in Ploidy Level (Haploids and Polyploids)

In the olive, to use homozygotic genotypes in the genetic improvement programme is of unquestionable importance. Considering the prevailing auto sterility and the juvenile phase that characterizes this species, the achievement of a homozygosis by auto-impollination seems very improbable, while the cultivation of macro or micro spores could help the achievement of such objective. Nevertheless, up to now, the regeneration of plants from these tissues is not possible. To the contrary, the production of polyploids is more easily attained by using chemical products (colchicine, orizaline, *etc.*) or by physical means (ionizing radiations). Triploid and tetraploid plants have been produced from the mixoploid cultivar of Leccino and Frantoio, obtained by irradiation with gamma rays. The tetraploid plants are recognizable because they present larger leaves. The triploids have been selected from seeds germination of the bigger fruits (arisen from diploid ovocell and haploid pollen) collected from mixoploid and tetraploid plants and later chromosomes count. Both types of plants are tested like rootstocks.

Somatic Hybridization

The somatic hybridation is useful for the production of somatic hybrids and for the production of cytoplasmatic hybrids (cybrids). In the olive, even though vital protoplast has been produced able to regenerate the cellular wall and subsequently microcalli, it is still not possible to regenerate complete plants from these cells. In case the regeneration succeeds, besides the production of hybrids plants could be obtained with high degree of variability as in regeneration from protoplasts.

Synthetic Seeds Production

Somatic embryogenesis can be useful for production of "synthetic seeds", that is, somatic embryos or apical buds encapsulated in a substrate of protection and/or nutrition, like sodium alginate to be used for micropropagation or for germoplasm conservation. Micheli *et al*, (1998) obtained good results with the cv Moraiolo after encapsulation in an alginate matrix, containing a nutritive medium, apical and nodal buds from micropropagated shoot cultures, they maintained up to 49 per cent viability and grew satisfactorily when stored at 4°C for no more than 45 days.

Germplasm Conservation

Germoplasm conservation is a need of the hour for any fruit crop species and genotypes in order not to lose interesting genotypes as a result of the novel tendency to cultivate few cultivars (more productive and more adaptable to most environmental conditions) in the new olive plantations. The conservation of seeds without doubt the simplest and less expensive technology, has had little sense in plants commonly propagated vegetatively which have a high degree of heterozygosis like olive species. For this reason preservation of the plant or its parts is indispensable. At present, field conservation is the unique option for the olive species, even though there are varied risks of biotic and abiotic nature. In alternative to the clonal olive trees, the *ex-situ* conservation, that exploits the cultivation *in vitro*, seems to be promising by the slow growth conservation and the cryoconservation. It offers enormous benefits: i) high speed of multiplication ii) requires very limited space, and iii) possibility to maintain pathogen free plants.

Slow Growth Conservation

It preserves the shoot under *in vitro* conditions in an ideal substrate able to slow down their growth at low temperature (about ±4°C). The two cultivars (Leccino and Frantoio), preserved in refrigerator on substrate OM without hormones, either in the darkness or at 8 hours of light photoperiod at 20 µmol m^{-2} s^{-1}, showed a good capacity of conservation for 8 months, but only in darkness (Lambardi *et al.*, 2000). Gardi *et al.* (2001) observed a different behaviour between among: cv Frantoio could be conserved at 6° C for 5 months showing 100 per cent survival and growth after conservation, while for stored shoot cultures of 'Ascolana tenera'and 'Moraiolo' at 6°C and in the dark the maximum growth potential only up to 2 months.

Cryo-conservation

Promising results have been recently obtained with the application of cryogenic techniques to the long-term conservation of olive germplasm by applying the direct immersion into liquid nitrogen (–196°C) or by using the vitrification solution.

Germplasm Characterization

The recent development of the molecular biology allows evaluation of the genetic variability in olive trees in an objective and precise manner across technologies that emphasize eventual genetic changes is of biochemical type (isoenzymes and allozymes), and molecular (RFLPs, RAPDs ISSR, AFLPs, SNIPs, micro satellites). Nevertheless, it is important to note that none of the methods used at present can be absolutely certain that genetic variations do not occur during the cultivation *in vitro* or the germplasm conservation.

Isoenzymes

The isoenzymes represent only a small portion of the genome and first attempts at the use of biochemical markers for understanding the domestication process in the olive started in the 80s by using isoenzymes extracted from pollen or from leaves, and they were able to distinguish the polymorphism between the cultivars.

The isoenzymes, being the product of the gene expression, are regulated by different factors such as environmental conditions, tissue of origin, and the development of the plant. These factors give back the isoenzymes few reproducible, giving back difficulties for their interpretation. However, these are highly polymorphic and co-dominating marker. In the olive, they were isolated from some genes codifying proteins and involved in some essential biochemical processes. The first gene isolated was the Stearoil-ACP (δ9) Desaturasi, the key enzyme in the trial of in saturation of the fatty acids of the membranes and of the lipids of accumulation. It determines the formation of the double connection that transforms the stearic acid in to oleic acid, the main constituent of the olive oil. Subsequently, other genes were isolated, like the RBCL (*Ribulose Biphosfate Carboxilase*), the gene for NADH dehydrogenase of the chloroplast and the gene of the crytochrome B5. Different codifying genes for a family of proteins that induce allergic symptoms in the man were isolated in different research. These proteins differ in length but all of them are expressed during the formation of the pollen (Cançado *et al.*, 2013).

This technology based on the electrophoretic discrimination of different molecular shapes of some enzymatic proteins facilitates and speeds up the work of the breeders in the evaluation of the germoplasm present for the genetic improvement of the species and for the certification of the vegetative propagate material. They are carried out on number limited genotypes that they have had common origin. Data banks are being created containing information on sequences isolated from RAPD fragments that do not correspond with other sequences and fragments found of DNA characterizing young and adult tissues (Cançado *et al.*, 2013).

Molecular Markers for the Genetic Study of Olive

The identification of olive accessions or varieties is traditionally realized through the analysis and comparison of morphological and agronomics traits, most of which are only available during the stages of flowering and fruiting. Because olive requires three to five years to begin fruiting, the use of this method does not allow the early identification of mistakes or segregating plants. In addition, the phenotype of plants can be significantly altered by environmental effects, such as nutritional and sanitary status, making it more difficult to identify specific varieties. Another drawback is the excessive number of olive varieties available, more than two thousand, which makes the morphological differences among many of them very small and difficult to distinguish. To overcome this limitation, the use of molecular markers is becoming routine for the precise identification of olive cultivars. Molecular markers can be an important tool to investigate the high genetic diversity results and significant alterations in economic traits such as oil content, fruit size, seed size and resistance/tolerance to biotic/abiotic stresses. Consequently, the accurate identification of olive varieties is fundamental for the optimum use of this crop. Due to outcrossing fertilization among cultivars in olive, makes this fruit crop a good model for molecular marker studies. The molecular markers in association with morphological characterization are already being employed to identify olive varieties more accurately beside their application in early stages of development, even in embryos and plantlets. It is not a destructive method and requires a very small amount of plant tissue. Therefore, the use of molecular markers has become a very attractive technique. Nevertheless, the most important advantage of molecular markers is the high level of genetic identification, which is directly related to the nucleotide sequence of the DNA, eliminating any chances of external interference caused by environmental variation.

Molecular markers based on the polymerase chain reaction (PCR), such as random amplification of polymorphic DNA (RAPD), ISSR and microsatellite (SSR), have been used frequently with the objective of identifying olive plants. These techniques require modest laboratory infrastructure and are relatively low-cost when compared with other techniques. However, the cost of the amplified fragment length polymorphism (AFLP) technique is average compared with the other techniques, although it does not requires the very complex infrastructure that is frequently used for genetic study in olive trees. The development of markers based on the sequencing of double-stranded DNA, such as single-nucleotide polymorphism (SNPs), has also been shown to be efficient for the identification of varieties or accessions of olive. The application of this class of molecular markers in olive studies has been limited, but this will change in coming years with the adoption of new high-throughput DNA sequencing technologies. Currently, microsatellite molecular markers are some of the most applied in olive variety identification studies. Microsatellite molecular markers exhibit high reproducibility, enabling information exchange, data bank assembly and result standardization among different laboratories and research groups. Many research studies have shown that a few microsatellite markers are sufficient to distinguish more than 100 olive genotypes.

Alien Gene Transfer

The low efficiency of both conventional and some older unconventional genetic improvement techniques (protoplast technique and somaclonal variation) seem to indicate that genetic transformation is the most promising technique to speed up the development of new superior cultivars. In order to apply this technique an efficient *in vitro* regeneration method and the availability of suitable genes are crucial conditions. The success of producing transgenic plants depends on the number of transformed cells and their ability to differentiate shoots and embryos. Common explants used in the olive for the transformation experiments are the leaf petioles, zygotic tissues and somatic embryos from in vitro grown cultivars because they are able to regenerate plants.

Genetic Transformation of Olive

This technology created new opportunities to improve plants in a more controlled way because it allows the direct introduction of selected genes obtained from any organism, including microorganisms, plants and animals, into the genome of the target plant (Perani *et al.*, 1986). Conceptually, genetic transformation is the controlled and stable introgression of genes directly into the genome of a target organism (Torres *et al.*, 2000). Therefore, genetically modified plants are plants that express one or few exogenous genes, which may or may not exist in the original genome and which are obtained by genetic engineering techniques (Gander and Marcellino, 1997).

Advantage of Genetic Transformation

- ☆ It allows plant breeders to overcome the sexual barriers between species, genera, families and even phyla, permitting the unrestricted exchange of genetic material among organisms.
- ☆ It is very useful to avoid the undesirable effects of genetic drag, which occurs when genes of interest are genetically linked to genes controlling undesirable traits since these genetic drag phenomena delays the recovery of the original genetic background.
- ☆ It can be utilized to improve cultivars while avoiding the need for slow backcrossing processes to maintain the original phenotype.
- ☆ Transgenic plants can be used as models in studies of gene expression and regulation; the role of genes in metabolism, biochemical and morphological pathways; and the response to diseases and pests, among many other uses.

The *in vitro* culture of plant tissue is considered an essential requirement for the genetic transformation of the majority of plant species however, the unavailability of an efficient technique for genetic transformation in olive has limited the widespread adoption of this technology in this species, principally due to the recalcitrant nature of olive for *in vitro* regeneration. Pérez-Barranco *et al.* (2007) reported advanced biotechnological approaches for genetic improvement in olive. He has developed an efficient system for the regeneration of young plantlets using somatic embryogenesis in root segments of mature zygotic embryos. In addition to an efficient system for DNA integration, the successful transformation of plants requires an efficient selection of positive events by permitting the regeneration of transgenic cells and inhibiting the proliferation of escapes, which are tissues that were not transformed but were able to survive even in the presence of the selective agent. According to Pérez-Barranco *et al.* (2009), in olive, these parameters of genetic transformation still require intensive optimization to make this species more responsive and allow the adoption of genetic transformation as a routine procedure. The high rate of generation of escapes operationally prevents any attempt at transgenic event production on a large scale, either for scientific objectives or for commercial purposes, because it considerably increases the cost associated with the genotypic and phenotypic analysis of events. There are different methods of transformation, and the choice of method depends on the species to be transformed, the type of explants to be used (callus, zygotic embryo, leaf tissue, protoplast, *etc.*), the regeneration capacity of the explant and the availability of the materials.

The plant genetic transformation approaches are categorised into two broad categories: direct and indirect genetic transformation. The direct genetic transfer of DNA is based on physical and chemical methods, such as biolistics (also known as particle bombardment), electroporation and PEG (polyethyleneglycol), while indirect genetic transformation of DNA uses the bacterium *Agrobacterium tumefaciens* or *Agrobacterium rhizogenes* as a vehicle of transformation. Results for the genetic transformation of olive have been limited. Lambardi *et al.* (1999) tested different devices and bombardment conditions for the transient transformation of somatic embryos of the olive cultivar Canino, without success. Pérez-Barranco *et al.* (2007) obtained transient expression of the *gus* gene in plants derived from olive embryonic cells by particle bombardment. Torreblanca *et al.* (2010) obtained a transformation frequency of 20 to 45 per cent with the *A. tumefaciens* system and observed stable expression of the *npt*-II (neomycin phosphotransferase) gene, which confers antibiotic resistance in leaves of genetically transformed olive. The genetic transformation of the olive cultivar Canino with the *rol* ABC genes (plant oncogenes of *A. rhizogenes*) meditated by *A. tumefaciens* has also been reported. This gene is responsible for the modification of vegetative growth patterns and stature of the plant. Olive also transformed with a gene that encodes the synthesis of osmotina, a compound involved in increasing resistance to abiotic stress. *A. tumefaciens can be used* to introduce the *rol* ABC genes in the cultivar Dolce Agogia, using leaf petioles as the explant for transformation. In addition to putative restrictions due to environmental effects (gene flow) and allergenicity, incorporating genetic transformation in olive breeding provides the attractive advantage of speed; the establishment of a new cultivar using traditional methods may take decades. The risks involved with this technology can be efficiently reduced or even avoided by the careful application of safety practices, thus potentially mitigating possible harmful effects to the environment and consumers.

References

Bao, Z-H., Ma, Y-F., Liu, J-F., Wang, K-J., Zhang, P-F., Ni, D-X., Yang, W-Q. 1980. Induction of plantets from the hypocotyl of *Olea europaea* L. *in vitro*. *Acta Bot Sin* **2:** 96-97 (in Chinese).

Cançado, M.D.A, Setotaw, T.A. and Ferreira, J. L. 2013. Applications of biotechnology in olive. Geraldo. *African Journal of Biotechnology* **12**(8): 767-779.

Gander, E.S., Marcellino, L.H. 1997. Transgenic plants. *Biotecnologia Cienc. Desenvolv* **1**: 34-37.

Gardi, T., Micheli, M., Piccioni, E., Sisani, G., Standardi, A. 2001. *Italus Hortus* **8**(4): 32-40.

Lambardi, M., Benelli, C., Amorosi, S., Branca, C., Caricato, G., Rugini, E. 1999. Microprojectile-DNA delivery in somatic embryos of olive (*Olea europaea* L.). *Acta Hortic* **474**: 505-509.

Lambardi, M., Benelli, C., De Carlo, A., Fabbri, A., Grassi, S., Lynch P.T. 2000 *Acta Hortic*. (Proc. "4th Int Symposium on Olive Growing – OLIVE 2000". Valenzano-Bari, Italy, 25-30 September.

Mencuccini, M., and Rugini E. 1993. *In vitro* shoot regeneration from olive cultivars tissues. *Plant Cell Tissue and Organ Culture* **32**: 283-288.

Micheli, M., Mencuccini, M., and Standarti, A. 1998. Encapsulation of *in vitro* proliferated buds of olive. *Adv. Hort. Sci.* **12:** 163-168.

Perani, L., Radke, S., Wilke –Douglas, M., Bossert, M. 1986. Gene transfer methods for crop improvement: introduction of foreign DNA into plants. *Physiol Plant* **68**(3): 566-570.

Pérez-Barranco, G., Mercado, J.A., Pliego-Alfaro, F., Sánchez-Romero, C. 2007. Genetic transformation of olive somatic embryos through biolistic. *Acta Hortic* **738**: 473-477.

Pérez-Barranco, G., Torreblanca, R., Padilla, IMG., Sánchez-Romero, C., Pliego-Alfaro, F., Mercado, J.A. 2009. Studies on genetic transformation of olive (*Olea europaea* L.) somatic embryos: I. Evaluation of different aminoglycoside antibiotics for nptII selection; II. Transient transformation via particle bombardment. *Plant Cell Tissue Organ Cult* **97**(3): 243-251.

Rugini, E. 1984. *In vitro* propagation of some olive cultivars with different root-ability and medium development using analitycal data from developing shoots and embryos. *Scientia Horticulture* **24:** 123-134.

Rugini, E., Caricato, G. 1995. Somatic embryogenesis and plant recovery from mature tissues of olive cultivars (*Olea europaea* L.) 'Canino' and 'Moraiolo'. *Plant Cell Rep* **14**(4): 257-260.

Torreblanca, R., Cerezo, S., Palomo-Ríos, E., Mercado, J.A., Pliego-Alfaro, F. 2010. Development of a high throughput system for genetic transformation of olive (*Olea europaea* L.) plants. *Plant Cell Tissue Organ Cult* **103**(1): 61-69.

Torres, A.C., Ferreira, A.T., Sá, FG., Buso, J.A., Caldas, L.S., Nascimento, A.S., Brígido, M.M., Romano, E. 2000. Glossário de biotecnologia vegetal. Editora Embrapa-CNPH, Brasília, DF, Brazil.

9

Climate and Soil Requirements

Olive trees like cool winters and hot summers (Mediterranean-like climate) to survive and produce crop successfully. A mature tree can survive temperatures down to -9.4 °C for a limited period but continuous exposure below that temperature can be fatal. Even though olives are evergreen trees, they also needed a cool winter rest to prepare for their main shooting, flowering and fruiting in the spring. For most varieties some winter chilling is essential. Throughout the world olives are grown in climates which range from very cold area of Tuscany (Italy) where minus 20°C is common, through to warmer areas of Seville (Spain), where some regions don't even reach 0°C during winter. The some of the olive genotypes can survive outside of the typical Mediterranean zone but unable to produce economically viable yield. The maximum and minimum summer temperatures are also govern the growth and development of fruit-bearing foliage. Most olive growing regions of the world have average maximum daily temperatures, in the hottest month of summer, somewhere above 30°C. Afternoon temperatures as high as 45°C have very little effect on mature olives as they have an inbuilt mechanism which temporarily shuts down their system until the cooler part of the day arrives.

Apart from the cool winter and warm summer requirements, the moisture levels of the tree must also be adequate. On an average olive crop require about 950 mm water during their entire growing season. A well distributed rainfall over the growing season is conducive for proper growth and development of trees. Moisture stress or prolonged drought during summers causes fruit drop and ultimately leading to poor yield. Inadequate or delay in rain during winter season delays emergence of new flush, lowering a substantial flower bud differentiation. In regions where the summers are cool or short, it is safer to choose early-ripening varietals that will accumulate enough oil in the fewer months of heat. Regions with colder winters should look for cultivars that come originally from mountainous or cooler regions. Regions which are very close to the ocean or experience a great deal of fog should lean away from varieties that are susceptible to leaf fungal infections.

In India, where the wild olive trees are growing are located in the North Western Himalayan states of Jammu and Kashmir, Himachal Pradesh and Uttarakhand adjacent to the naturally flowing rivers of hilly areas. The hills and mountains are stretched from the East to the West along with the Himalaya. Although it is located in between 28° 43′ to 36° 10′ North latitude, is mainly a temperate region, especially the Northern parts, due to the higher altitude and the presence of close Himalayan range. The Northern sides of the mountains are cooler than the Southern ones. It has diverse climatic conditions with many micro-climatic pockets.

It receives plenty of rain from the Bay of Bengal of the Indian Ocean during summer (monsoon) and less rain from the North-West side in the winter. There are some areas in these states which are suitable for olive growing. This has proven also by the presence of wild species of olive at altitudes ranging from 1000 to 2000 meters above sea level (asl.)

Effect of Climatic Factors on Olive Cultivation

Frost

Occurrence of autumn frost is more harmful, since it affects fruit maturation and oil accumulation. Hail storms at pre-bloom, bloom and post-bloom period causes higher abscission of flowers and young fruits. At blooming stage hail storms render trees completely devoid of fruits, inducing alternate or irregular bearing in its trees. Mature olive varieties can survive and crop well even in the very cold areas of olive growing regions. Some varieties will also fruit well in 'no frost' areas as long as the winters are cool enough. Short cold snaps of down to minus 10°C are often not overly detrimental to mature olive trees with sufficient moisture in the soil to avoid stress. Some varieties can survive upto minus 15°C as long as they are in good health. Most olive trees are killed minus 20°C (at ground temperature) such can occur in the coldest olive growing regions of the world. An important aspect to be considered is the locality temperatures during planting time, which should not fall below minus 5°C. Because very young trees can tolerate low temperatures (minus 5°C) when the good tree health and sufficient moisture are available.

Temperature

In India, it can be planted from 800-1400 m above mean sea level. Temperature is the most important factor influencing its cultivation. Olive cultivation require a temperature range of 7-35 °C; however, 15-20 °C ideal. Occurrence of prolonged hot and dry summers coupled with acute water stress cause post anthesis flower and fruit drop. Adequate chilling at 7.2 °C during winter is essential to break rest period and to promote fruitfulness.

Sunlight and Exposure

Olives require full sunlight. An evergreen tree, the olive needs sunlight year-round and not just in summer. Therefore, tree should be planted in an exposed area where no shading occurs. The trees are also wind tolerant and able to withstand heavy winds with no damage. The sculpted look of many mature olive trees comes from being planted in exposed areas.

Moisture

Olive in general requires low water but regular access to moisture leads to better fruit production. Olives may rarely require irrigation in climates with moderate rainfall, or approximately 1 inch of rainfall per week. The trees do not tolerate soggy soil; so, areas with heavy rainfall or regular flooding are not suitable for olive trees. In drier climates, regular irrigation year-round makes up for the lack of natural moisture.

Protection Moisture for Olive Trees

Olive trees are warm-temperate plants that can survive a light freeze without protection. Even a hard freeze, down to 23°F, is not a problem, as long as the lowest temperatures last for only a few hours. At 22°F, olive trees can suffer damage to small branches, especially new growth. This superficial tip burn does no permanent harm to the tree.

The selection of site for planting olive is also important. Olive trees in cold winter areas should be planted in the lee of the coldest winds. Heavily wooded areas to the north are good. Small-stature cultivars such as 'Arbequina' can be planted on the south side of a house for wind protection. Paved areas near the tree help retain some heat. Large bodies of water also provide some protection, especially if they are deep. Covering of trees should never be done with plastic sheeting. This does little or nothing to stop the transfer of cold and, if left on until the sun is high and warm, can burn the tree. Implementation of the following guidelines will help with the overwintering of olives in cold areas.

Care for Container-Grown Trees

- Leave container-grown olive trees outside in a sunny location until temperatures fall below freezing.
- Before bringing trees indoors during the winter, check them for insect pest and diseases. It is a good idea to treat with a suitable chemical because pests can thrive in the indoor conditions.
- Olive trees require strong light. Consider using grow lights for trees that will be indoors for long periods.
- When sub-freezing temperatures are expected, bring potted olive trees indoors to a cool, well-lighted place. A room with a south-facing window and a temperature of 40-50° is ideal.
- Move it in and out. When temperatures are above 40°, move your tree outside to a sunny location during the day. Move it back inside if freezing is expected overnight.
- Keep the soil slightly moist (but not soggy-wet) through and through. Check the top of the soil and drain holes to be sure soil is not just damp on the top.

Soil

Olive can grow well in a wide variety of soils, but for optimum growth and productivity it requires deep, fertile and well-drained soil. It prefers non-stratified, moderately fine textured soils, including sandy loam, loam, silt loam, clay loam, and silty clay loam. Such soils provide aeration for root growth, are quite permeable and have a high water holding capacity. Sandy soils do not have good nutrient or water holding capacity. Heavy clay soils often do not have adequate aeration for root growth and will not drain well. Olive trees are shallow rooted and do not require very deep soils to produce well. Soils 4-6 feet deep are well suited. Soils having an unstratified structure of four feet are suitable for olives. Stratified soils, either cemented hardpan or varying soil textures within the described profile, impede water movement and may develop saturated layers that damage olive roots and should be ripped. Olives tolerate soils of varying chemical quality. On the contrary, olive trees do not tolerate wet soils for a prolonged period, since it results in eventual death of its roots. Soil pH of 6.5 -7.5 is ideal. However a pH exceeding 8.5 adversely affects its growth and productivity. Its trees can tolerate a fairly high amount of Ca and B.

Steep Slope Soils

Olives can be grown on steep inclines or terraced, although access and mechanical harvesting, spraying, and other orchard maintenance chores can be difficult and more costly. There has been debate on whether terraced trees should be placed on the edge of the terrace or against the hillside. Inclines greater than 20° are inappropriate for most tractors and other vehicles. Inclined properties cannot be assumed to have good drainage.

10
Varietal Wealth

Olive is very rich in varietal diversity in spite of the disturbance of the environments where they are cultivated. The varietal diversity can be utilized in olive improvement programme and farmers can grow varieties as per their needs and demand. As per reports published by FAO on Plant Production and Protection Division Olive Germplasm (FAO, 2010) that olive has more than 2629 different varieties, with many local types and ecotypes. The problem of olive germplasm classification is not only complex due to the richness of its genetic patrimony, but also due to lack of reference standards. There is a great confusion existing regarding the cultivar names with numerous cases of misnomers, synonymous and duplication.

On the basis of utility, the varieties are classified as:

1. **Oil type cultivars:** Leccino, Carolea, Ottobratica, Biancolilla, Zaituna, Moraiolo, Coratina, Pendolino, Cipressino, Frontoio, Maurino, Canino, Carnicobra Attica, Crisolia.
2. **Pickle and table type cultivars:** Tonda Ibea, Picholine, Kalamon, Gordal Conservolia, Manzanilla, Itrana, Kalamon, Thrumbda, Ascolano, Coratina, Hojib Lanca, Mission.
3. **Cultivars grown in India:** Coratina, Leccina, Frontoia, Pendolino, Carolea, Tonda Ibea, Cipressino.

Major Olive Cultivars Developed and their Characteristics

A large number of the olive varieties has been developed by many countries viz., Albania, Algeria, Argentina, Australia, Azerbaijan, Brazil, China, Cyprus, France, Greece, Egypt Iran, Israel, Italy, Japan, Jordan, Montenegro, Morocco, Nepal, Portugal, Slovenia, South Africa, Spain, Tunisia, Turkey and USA. Globally, Olive germplasm Bank of Spain has over 350 cultivars collected from different countries is one of the largest collections of the olive germplasm (Caballero *et al*, 2008). A database on olive germplasm includes about 1,250 cultivars from over 50 countries at has been developed by the National Research Council of Italy and Food and Agriculture Organisation (FAO) of the United Nations (Caballero *et al*, 2008 and Grati *et al*, 2006). Some of the very popular olive cultivars of the world have been given as below.

1. Aggizi Shame

Originated from Fayoun, an Egyptian variety, self-compatible and has medium pistil abortion. Higher productivity, drought hardy and moderately susceptible to olive flies. Fruits are large, freestone and has high flesh to stone ratio. Flesh is tasty, firm, and resistant to bruising. It has low oil content (7-9 per cent) and suitable for the production of green or stuffed olives.

2. Aggizi Oshime

Originated from Fayoun, an Egyptian variety has a high rooting ability and fruit productivity. Fruit mostly picked for green pickling. Which are characterised by large with medium flesh-to-stone ratio.

3. Aglandau

Originated in France, moderately productive, start bearing at intermediate age, oil content is medium, tree has medium vigour with spreading crown, usually self sterile; Picholine and Cayon are pollinizers. It is considered a hardy variety, with good cold tolerance but only moderate to drought tolerance. It yields a high quantity of excellent quality oil with good keeping qualities. This variety is quite bitter and is better suited to harvest when fairly ripe.

4. Amfissa

It is a Greek table olive grown in Amfissa, Central Greece. The tree has strong vigour with a spreading habit and medium dense canopy. The leaves are elliptic lanceolate shape. The harvesting time depends on its use which may be for green or black olives or for oil. Oil content is medium.

5. Arbequina

Originated in Spain. Early ripening variety and fruit does not turn black until overripe. A low-vigour variety with spreading, somewhat droopy growth habit. Early bearing and low vigour makes it fit for high-density plantings. Considered cold tolerant and adapted to a wide range of soil types with moderate tolerance drought and salinity. Usually self-fertile, but the presence of a pollinizer can increase yields. Tolerant of leaf spot by some sources, susceptible to olive fruit fly, low tolerance to black scale, moderately susceptible to olive knot and susceptible to *Verticillium*. It is one of the varieties with the highest percentage of oil extraction (20.5 per cent). The oils colour is green and yellow with a fresh and fruity scent depending upon agro-ecological condition. The oil is having no sting, very soft, light, sweet, delicate, bitter almond flavour and aroma of ripe fruit (Verma, 2006).

6. Arbosana

Originated in France, semi-dwarf cultivar. Small fruit, self-fertile and moderate cold tolerance. Trees get 12 to 15 feet tall and the fruit can either be pressed for oil or processed.

7. Azapa (Azapena and Sevillana de Azapa)

Fruit size varied greatly and ranged 8.0 -10.0 g. The fruit is typical elongated shape, point end, very fleshy with a thin skin. Derives its name from the Chilean valley in which it constitutes about 90 per cent area. Suitable for warmer regions experiencing no frost with a slight tendency toward alternate bearing. It is the most highly rated cultivar in Chile, South America, considered a good table olive.

8. Alfonsos

It is considered as one of the best olives. The fruits of this variety are black in colour and one of the best quality types. These are having smooth skin and have a distinct robust taste great for antipasto salads.

9. Balady, syn: Nabala Balady

Originated in Egypt and Palestine. Trees moderately vigorous, spreading canopy, fruits medium in weight, elongated shape, asymmetrical, pointed apex, full bloom period from mid April-late April, and matures in October-November. Tolerant to cold and drought, moderately resistant to olive fly, and higher yield. Fruit weight 2.8 g, pulp-to-pit ratio 3.9 and oil content 50.7 per cent (Verma, 2006).

10. Barnea

Bred in Israel, is a dual-purpose, disease-resistant cultivar, producing a generous crop. Trees are vigorous and erect, accommodating about 400 trees/ hectare. It responds well to irrigation. Considered susceptible to a number of leaf and root. Potential for heavy, early cropping. Medium to high oil content which has a strong flavour.

11. Barouni

Barouni is of North African origin. The fruit is quite large, ranged from 9-11 g, with a pit to stone ratio as Sevillano. Barouni is marketed as "Queen" olives in South Africa.

12. Belice

Cultivar from the Valle del Belice area of South-Western Sicily. It is a dual-purpose olive, grown both for oil and for the table.

13. Bosana

The most common in Sardinia and used primarily for oil, but is also delicious to eat with fruity, bitter, and spicy taste. Traditionally it is used for oil, and gives a good yield (17–18 per cent). The quality of the oil is improved if the olives are harvested early, at the start of the maturation. The fruits of greater size are often transformed to table olives, whether green or black.

14. Bouteillan

It is exclusively grown for oil and has a high oil yield. It is a hardy plant which need light but frequent watering and respond to canopy management. It growth is very fast and produces high yield on sustainable basis. Ripening time is intermediate. Highly self-compatible, pollinizers are Grossane and Cayon.

15. Cailletiere

Also known as Niçoise, which signifies its curing method, an important ingredient in the Niçoise salad. The fully ripe olives have "a dark colour that ranges from black brownish-purple to brownish-black. It is considered a productive cultivar, but with a tendency towards biennial bearing. Generally considered as self-fertile.

16. Canino

The tree is tall with an upright shape and compact crown. It is a typical oil (15-16 per cent) variety and has small fruit (1-2 grams) and spherical in shape. At harvest, the olives are never all black because the maturation is late (December) but scattered out. Recommended pollinators: Olivone, Frantoio, Pendolino and Leccino.

17. Cerignola

Native to Italy. Tree is medium to low vigour, erect or spreading with low tolerance to cold, susceptible to drought. It is a demanding variety, requiring good cultural practices. Partially self-compatible, often grown with pollinizers such as Mele, San Agostino and Termite di Bitetto. Popular because of large size of fruit, which used for green olives in brine and as black table olives. Fruit ripens early and the flesh is clingstone and fibrous with high flesh to pit ratio.

18. Chemlali

Native to Tunisia. A very large tree reaching even more than 40 ft., thrive in hot summers and can be suitable for coastal regions also. Fruit yields sweet and smooth oil with a bit of almond flavour, like Arbequina. This is a common variety used in Tunisian and Moroccan olive oils (Verma, 2006).

19. Cipressino

Also known as Frangivento. This has vigorous vegetation, rapid upward growth with a typical rising habit and gathered crown. The leaves are lanceolate, medium-small, flat surfaced and dark gray green. The fruits are rounded oval shape, at maturity having weighs 2-5 grams and are used for producing an oil yielding 15-17 per cent that is fine and light. Maturation is spread out but is complete between the middle of November and the middle of December. Beside this it is a self-sterile variety of excellent and constant fruiting with a notable percentage of aborted ovaries (50-60 per cent). Pollinators: Frantoio, Leccino, Pendolino. Excellent resistance to salty winds, good resistance to climatic extremes and parasites

20. Columella Syn: Colombale, Columello

Native to France having medium tree vigour with intermediate and alternate production habit. Exhibits low to medium resistance to cold can give good consistent yields in a warm location where it is grown in combination with other cultivars. Moderately susceptible to peacock spot and highly susceptible to *verticillium*. Green harvested fruits yield aromatic, sweet and fruity oil. Ripening is medium early; fruit is fragile and prone to bruising.

21. Coratina

Native to Italy, high oil yielding variety with cold tolerance characteristics. It is highly self sterile, productivity is high and starts of bearing early but ripens late. Tree is medium vigorous, dense, spreading growth habit with very good adaptability to different soils and climates. Adaptable and tolerant of cold and moderately tolerant to drought and salinity. Fruits are large, elongated ovals and asymmetrical in shape. Yields high quantity of good quality oil which is exceptionally high in polyphenols, resulting in excellent stability. Coratina requires a long growing season and heat to ripen to produce good oil (Verma, 2006).

22. Cornicabra

Also called Cornezuelo, Corniche or Osnal. They all refer to the horn shape of the olive typically long, slightly curved and asymmetrical. It weighs on average about 3.0 grams, with a flesh-to-stone ratio of 5:1 and an oil yield of 19 per cent. The ripening period is late but long, from the end of October to the beginning of January. It has a golden colour with greenish tones and a fruity flavour. When obtained from mature fruit picked toward the end of the harvesting period, its flavour and texture are more similar to tropical fruits.

23. Dolce

A French cultivar used for oil making and processing. Tree vigour medium (3.15 m), bloom from March-April, higher yield 25 kg/tree), medium fruit weight (3.5 g), high flesh-to-stone ratio, medium oil content (33.5 per cent), elongated fruit and elongated stone shape.

24. Dritta

It is a variety of olive tree typical of Aprutino Pescarese in the province of Pescara (Abruzzo). The fruit used to produce extra virgin olive oil with excellent chemical and organoleptic properties.

25. Empeltre

Native to Spain and also known as Aragonesa, Injerto, or Mallorquina. It is a dual-purpose cultivar. The fruit is long, asymmetrical and slightly bulging on the back, weighing about 2.7 grams with flesh-to-stone ratio of 5:3. It has a relatively low but acceptable oil content of 18.3 per cent. Oil is smooth, pale yellow colour, sweet and aromatic, with no bitter taste, ideal for blending with stronger oils.

26. Frangivento

Very vigorous tree nature with rapid upward growth. The fruit rounded oval in shape, weighing 2-5 grams at maturity, yielding 15-17 per cent fine and light oil. It is self-sterile; varieties like Frantoio, Leccino, and Pendolino can be used for pollinizer purpose.

27. Frantoio

Syn: Razzo, Corregiol, Paragon Frantoiano, Correggiola, Correggiolo Gentile: Native to Italy, dual purpose variety. Trees are moderately vigorous with spreading-drooping growth habit and medium-dense canopy. Yield fairly highly and consistently, high adaptability and cold sensitive. Frantoio is self fertile, but using pollinizers increases yields. Pendolino, Leccino and Maurino can be used as pollinizers, it is susceptible to peacock spot, scale insects and olive knot. It is also moderately susceptible to olive fly and to *verticillium* wilt. It is world's premier oil varieties with high oil yield. Frantoio is a late ripening variety.

28. Gaeta

Italian black olive which is cured in dry salt and some oil which is usually packed with rosemary and other herbs. Have a mild flavour and a wrinkled look.

29. Gemlik

It is a variety from the Gemlik area of Northern Turkey. They are small to medium sized black olives with high oil content. This type of olive is very common in Turkey and is sold as a breakfast olive in the cured formats of Yagli Sele, Salamura or Duble, though there are other less common curings. The sign of a traditionally cured Gemlik olive is that the flesh comes away from the pit easily.

30. Hamed

Egypt. This variety is tolerant to drought, salinity and high temperature. It has a good rooting ability and its start of bearing is intermediate. It is self-compatible and has a low pistil abortion rate. Its productivity is high and constant. The fruit is large and very sensitive bruising during transportation and handling. The flesh-to-stone ratio of the fruit is high. Freestone, it is used for green and black pickling.

31. Hardys Mammoth

The source of this cultivar is unknown. It is a tough cultivar that bears good early season crops, and larger size fruits suitable for pickling. The oil content is high.

32. Hojiblanca

Popularly known as Casta de Cabra or Lucentino, The name comes from the white colouring on the underside of the leaves. The olive is large, up to 4.8 grams, and is spherical in shape. The ratio of flesh to stone is 8:1. Maturation takes place from the end of November to the end of December. The oil yield is relatively low, about 17-19 per cent with high levels of fatty acid (75 per cent) and linoleic acid (7 per cent).

33. Itrana

Dual purpose and constant high yielder variety. It is most commonly pickled when black; the fruit is large and freestone. It gives a medium yield of oil that is highly regarded for its fruity character. This cultivar is late to ripen and the tree is self sterile and should be planted with pollinizers such as Leccino and Frantoio.

34. Kalamata

This is the second most important olive variety in Greece. A large, black olive with a smooth and meat like taste, is named after the city of Kalamata, Greece, and is used as a table olive. It is mostly grown in the southern region of the Peloponnese (Kalamata, Lanconia) and also in Central Greece. The fruit is of medium size (from 3 to 6 g average) and has one curved side. It is cylindroconical in shape. The skin turns black on maturity and the flesh retains a good texture. The tree is moderately vigorous, grows upright and has characteristically large leaves. The flesh to stone ratio is good – about 8:1. The processed olives are of excellent texture, colour, taste and aroma. Kalamata olives are in great demand worldwide.

35. Khalkidhiki

This is a large Greek green olive that is oval in shape and commonly harvested while still young. Cured in brine and has a firmer meat texture that provides a soft yet full flavor which is a little tart and with some hint of peppery taste.

36. Koroneiki (Coronaiki)

Native to Greece and characterized as dual purpose variety for table and oil, self-fertile, precocious with medium and oval fruit shape. Tree is dwarf in nature about 10-15 ft. Fruit very small (1-2 g), slightly asymmetric, ripens early, with quite high oil content of exceptionally good quality. Leaves are silvery green. The chemical characteristics of Koroneiki oil are excellent. The oil is high in oleic acid and very stable. Fruit yield is high and regular. The small fruit size renders it difficult for mechanical harvesting. Ripen early with high and constant yields. Performs well under plains, lower hillsides and coastal areas where the climate is relatively warm. Good drought resistant variety but cannot tolerate cold well (Verma, 2006).

37. Leccino

It is a rustic variety which is widely planted throughout the world. The tree is quick to produce and resists well adverse climatic conditions and parasites. More recently it has been used for the production of table olives, semi-ripe or black. This variety is self-sterile and so needs a pollinator, principally Pendolino or Maurino.

38. Lechín

It is a vigorous variety; name corresponds to the white colour of its flesh and its oily liquid. Fruit weight an average of 3 g, and have an asymmetrical ellipsoidal shape. However, the oil content (fatty acids) is not very high (18 per cent) corresponds to a flesh-to-stone ratio of between 7:2 and 8:5. Tolerant to cold winters and poor soils. Has high content of palmitic acid (12-13 per cent) and reduced content of stearic acid low saturated acid. Nevertheless this combination produces unstable organoleptic characteristics, with a tendency to oxidation. The flavour is slightly bitter, leaving an after taste of green almonds. This oil is ideal for tapas and sweets.

39. Liguria

Italian variety of olives that have a stronger flavour. They are usually packed in brine and salt with the inclusion of stems.

40. Lucques

It is found in the South of France (Aude département). They are green, large, and elongated. The stone has an arcuated (bow) shape. Their flavour is mild and nutty.

41. Lugano

A very popular Italian black olive variety which exudes a strong salty flavour. Preserved together with the leaves in brine with some herbs.

42. Maalot

(Hebrew for merits) is a disease-resistant, Eastern Mediterranean cultivar derived from the North African Chemlali cultivar in Israel. The olive is medium sized, round, has a fruity flavour and is used almost exclusively for oil production.

43. Manzanilla

One of the most popular table olive cultivars internationally. It is a robust variety with a well-developed canopy. It sets single fruits of medium size that are symmetrical and apple shaped, as indicated by the name (Manzanilla – small apple). Low to moderate in vigour, spreading growth habit and medium density. Alternate bearing, self fertile but pollinizers improve yields. It is moderately tolerant to both cold and drought. Fruit is round and freestone, with a high ratio of flesh to pit. It is harvested green for curing and can be picked when changing colour. Manzanillo can also make good oil with distinctive varietal character. The oil is difficult to extract when fruit water content is high. It is very susceptible to verticillium wilt and olive knot. It is attractive to olive fly and moderately susceptibility to scale insects and peacock spot.

44. Maraki

It was originated from Siwa Oasis, Egypt and main oil producing cultivar of Egypt. The fruit is very heavy and has a medium flesh-to-stone ratio. Its oil is very high in oleic acid content and is of medium bitterness. Its rooting ability is low while its crop intensity is very high. The best time to harvest this variety is from November to December.

45. Maurino

Originated at Italy. Tree medium in vigour with erect growth habit. Moderate or high cold tolerance. It is a good pollinizer, producing large amounts of fertile pollen compatible with many cultivars. Pollinizers for Maurino include Pendolino, Leccino, Moraiolo, Frantoio, Lazzero and Grappolo. Fruit is uniform in ripening; the oil content of fruits is medium to very good, and good in quality. It is considered resistant or moderately resistant to leaf spot and moderately susceptible to olive knot.

46. Mavroelia (Greek black olive)

A small black olive from the Messinia district of Greece, the tasty fruit is picked when ripe but well before over ripening begins and before the olives spoil and shrivel from frost. They have to be transported as quickly as possible to the processing plant where they have to be sorted, washed and immersed in a brine solution.

47. Memeli

A Turkish olive used for split green olives, green olives in brine, black olives and olive oil.

48. Mission

It is by far the most popular cultivar grown in South Africa. It is a dual-purpose olive (can be used for oil and table use) and is very suitable for black processing and gives reliable production under a wide range of conditions. It originates from California, USA. The fruit is of medium size with a fair flesh to pit ratio, which improves as the fruit ripens. Mission is most suitable for black table olives and will give a product an attractive appearance, medium firm to plump texture and good taste when processed correctly.

49. Moraiolo

Origin at Italy. Trees are low to medium in vigour with upright growth habit. It has low tolerance for cold. It is self-sterile and incompatible with Leccino; pollinizers include Pendolino, Maurino, Lazzero and Rosino. Gives high and regular yields and is intermediate in ripening time. It yields high content of excellent oil, which is easy to extract. Susceptible or highly susceptible to leaf spot, susceptible to olive knot and *verticillium*.

50. Nabali

Originated in Palestine, known locally as Baladi, which, along with Souri and Malissi, is considered to produce among the highest quality olive oil in the world.

51. Nafplion

It is a small green olive grown only in the Argos valley in Greece. Nafplion olives are traditionally cracked and cured in brine.

52. Nicoise

French small black olive, harvested only once fully ripened. These olives have a larger pit and are often preserved with the stems and a variety of herbs. The Nicoise olive exudes a smooth nutty flavour.

53. Pendolino

Native to Italy. Moderate in vigour with dense drooping growth habit. Flowering early and profuse with long bloom period which makes it an ideal pollinizer. Self-incompatible, fairly tolerant to cold temperatures. Fruit small, delicious green and black table olive. Maturity is mid season. Susceptible to olive knot and peacock spot, it is highly susceptible to verticillium and sooty moulds, moderate susceptible to the olive fly.

54. Picholine

Picholine are green olives of French origin. It is partially self fertile, Manzanillo and Leccino are good pollinizers. Resistant to leaf spot, moderately resistant to olive knot and verticillium wilt, with medium level of drought and cold tolerance and attractiveness to olive fly. Freshly brined right after being picked in citric acid has a subtle and mild salty taste.

55. Picual

Originated at Spain. Trees have high productivity and named for its pointed tip (pico). The fruit is medium to large in size (3.2 g), flesh:stone ratio 5:6. Tree is spreading, medium in vigour, cold tolerant and highly adaptable, tolerant of salinity and excess soil moisture, but sensitive to drought and calcareous soils. Flowers in mid-season and the cultivar are usually self-fertile.

56. Picudo

Named for its highly sweet oil that birds peck at the fruit at the time of ripening or from its shape, a curved pointed tip with a marked nipple. Also called Basta, Carrasqueño, Paseto or Pajarero. Picudo is the second-largest cultivar used for oil extraction, weighing an average of 4.8 g. The flesh: stone ratio is 6:3. The average oil yield is about 20 per cent with 15 per cent linoleic acid and up to 65 per cent monounsaturated oleic acid, with a tendency to oxidation. The flavour of the oils is soft, with an exotic fruit after taste. These fruits are excellent as table olives, green and black.

57. Ponentine

Italian variety of black olives. Usually preserved in brine with some salt. These black olives have a milder flavour.

58. Santa Caterina

Originated at Tuscany, Italy. Very large fruit with a high flesh to stone ratio. Needs a pollinator. Early in fruit bearing and spreading nature tree habit.

59. Sevillano or Gordal

This is a Spanish cultivar with large fruit, an average of 100 – 120 olives per kilogram, with a flesh to stone ratio of 7,5:1. It is slightly heart-shaped and the epidermis is thin and speckled with white spots. The only drawback of this variety is that the flesh is difficult to detach from the stone.

60. Spanish

Spanish green olives are the most popular green olives around the world. Brine and fermented in some acid solution for a good period of 4 to 6 months. The long processed of fermenting gives it a unique salty flavour which makes the Spanish green olives the top of its class.

61. Taggiasca

It is the dominant variety in the Italian region of Liguria. Trees vigourous, drooping growth habit, adapts well under coastal and hilly terrain, considered sensitive to cold, particularly in the spring, and also to drought and wind. It is susceptible to olive knot and leaf spot. Productivity is high and constant with high content of easily extracted oil. Traditionally, this variety is harvested late when it is very ripe, and it yields the fruity and delicate oil. The quality of the oil is excellent. Taggiasca is considered a late bloomer and late-ripening variety with alternating crops annually.

62. Toffahi

It originates from Fayoum district of Egypt. It is self-compatible and has a low pistil abortion rate. Flowering and harvesting are early. Its productivity is medium. The fruit is freestone and has a very high flesh-to-stone ratio; it is used primarily for green pickling. It is moderately sensitive to damage during transportation and handling. Although it changes colour early, it is not suitable for black pickling because the fruit has a tendency to become over soft and to ferment during the process. The fruit is large and its oil content is low (5–7 per cent). It is drought hardy, and moderately sensitive to olive fly.

63. Verdale, S.A. Verdale and Wagga Verdale

This cultivar native to France. The tree has medium size with drooping branches. They vary significantly in size of fruit, oil content, flesh-to-pit ratio, harvesting times, hardiness *etc.* The South Australian Verdale, a larger, oval fruited selection of the normal Verdale weighs approximately 7-10 g, yielding 100-140 olives/kg. The Wagga Verdale reportedly has smaller fruit, similar to the plain Verdale, but a heavier crop than the South Australian Verdale. The seed is too large compared to the other table cultivars. The fruit, although large pitted, are used for table olive processing which results in a pleasant tasting, good textured olive. The oil content is generally reported as low (7-23 per cent) generally towards the lower end).

64. Verdial

The name Verdial is given to a number of local varieties with similar characteristics. A common characteristic being the thickness of the skin of the fruit causing higher triterpenic alcohols in the oils, especially in the virgin olive oils. The fruit is large with variable oil content of up to 22 per cent, flesh: stone ratio is also high, which makes it an ideal dual purpose variety. Oil is extremely sweet taste and no bitter with a pleasant taste having high content of linoleic acid, which make it somewhat unstable and require protection from heat, light and air to be preserved perfectly.

65. Wateken

It is ancient variety comes from the Siwa Oasis, Egypt. The average weight of the fruit is very high and it has a medium flesh-to-stone ratio. Although dual purpose (oil and picking), it is mainly used for oil production, giving an oil which is high in oleic acid content and low in bitterness. The best time to harvest the fruit is between October and December.

66. Zatuna

Native to Greece, green colour variety. Mainly used for oil purpose and oil content ranges from 20-25 per cent. This is good pollinizer for Pendolino, Cipressino and Coratina.

Table 5: Major Olive Cultivars of World and their Characteristics

Cultivars	*~ per cent Oil*	*Cold Hardiness*	*Fruit Size*	*Polyphenol Content*	*Pollenizer Varieties*
Aglandau	23-27	Hardy	Medium	Medium	Self compatible
Arbequina	22-27	Hardy	Small	Low	Self compatible
Arbosana	22-27	Hardy	Small	Medium	Self compatible
Barnea	16-26	–	Medium	Medium	Self –Manzanillo-Picholine
Bosana	18-28	–	Medium	High	T de Cagliari-Pizze Carroga
Chemlali	26-28	–	Very small	High	Self compatible
Coratina	23-27	Hardy	Medium	Very high	Self -ogliarola
Cornicabra	23-27	Hardy	Medium	Very high	Self compatiblo
Empeltre	18-25	Sensitive	Medium	Medium	Self - Arbequina
Frantoio	23-26	Sensitive	Medium	Medium high	Pendolino - Leccino
Hojiblanca	18-26	Hardy	Large	Medium	Self compatible
Koroneiki	24-28	Sensitive	Very small	Very high	Mastoides
Lechin Sevilla	22-23	–	Medium	Medium	Hojiblanca - Picual
Leccino	22-27	Hardy	Medium	Medium	Frantoio - Pendolino
Manzanillo	15-26	Sensitive	Large	High	Sevillano - Ascolano
Moraiolo	18-28	Sensitive	Small	Very high	Pendolino - Maurino
Picudo	22-24	Hardy	Large	Low	Hojiblanca - Picual
Picual	24-27	Hardy	Medium	Very high	Self - Picudo
Picholine	22-25	Moderate	Medium	High	Self - Aglandau
P. Marocaine	22-25	Hardy	Medium	High	Self – P. Languedoc
Taggiasca	22-27	Sensitive	Medium	Low	Self compatible
Verdial Huevar	24-26	Hardy	Medium	High	Manzanilla – Gordal

Sources: cesonoma.ucdavis.edu

Table 6: Olive Cultivars Introduced in India from different Countries through NBPGR, New Delhi

Country	*Varieties/Cultivars*
Portugal	*Olea europaea* var. *silvestris*
Spain	Cresent olive, Lecpin, White whitlow – wort, Pin oak olive, Verdial, Farga, Picual, Blackish, Cornicabra, Manzanillo, Mission,
USA	Sevillano, Manzanillo, Criolia, Mission, Bidhel, El Hammam, Picholine du, Anguedoc, CucciAdrouppa, Gaidourelia, DMOR00 15, DMOR00 59, DMOR00 125, DMOR00137, DOLE137
Philippines	Nevadillo Blanco, Corregiolla, Regalaise, languedoe, Maerocarpa, Lucca, Attica, Dr. Fiac, Barouni, Big. Spanish, Mission, Verdale, Tarascoa, Bouquittier, Manzanitto No. 2, Manzanitto No. 14, Atro Rubens, Boutillon, Belle d' Espangne
Chile	Azepa, Ascolana, Seviliana
Tunisia	Chetoui, Chemlali
Argentina	Paraguay-580
USSR	No. 1351, Askolamo, Ackoiaro
Greece	Kalamon, Chondralia chalkidikis, Amfissis
Iran	Roghany, Gloleh, Fishomi, Mari, Shangeh, Zard
Israel	Barnea
Egypt	Toffahi, Aggizi Shame, Hamed, Balady, Maraki, Khouderi, Arbequin, Manzanillo, Picual, Koronaiki, Coratina, Frantoio, Kalamata, Dolce
USA	Arbequina, Ascolano, Ascolano Dura, Balady, Bouquetier, Bouteillon, Dole 0090, Franklin, Frontoio, Grossa Di Spagna, Kadesh, Leccino, Manzanillo, Mission, Mission Leiva, Nevadillo, No. 12 Sevillano, No. 31 Sevillano, No.1 Sevillano, Piconia, Picual, Rubra, Sevillano

Source: NBPGR Annual Report, 2013

There is certain procedure while selecting cultivar or its groups for a specific geographic area (ecotype). These are as followes:

- Selection should be made based on different growth habit for higher productivity and production efficiency.
- For high density orcharding dwarf characteristics cultivars should be selected.
- Vigorous cultivars should be selected for well spaced plantations which give a high production per tree and are better for mechanical harvest.
- Cultivars having up-right growth are suited for "monocono" training form, but for square or rectangular plantation cultivars may be different.
- Select three or more cultivars for a single block to effect pollination and production may be affected by one or more cultivars due to unfavourable climatic conditions.
- The indigenous olive cultivars belonging to the same area as cultivars belonging to other areas may suffer due to poor growth and pollination problems.

- ☆ Cultivars selected should help each other in cross pollination. If a single cultivar is planted in an orchard it needs 20 per cent pollinizer. Pollinizer is required for self-fertile cultivars also because they give better yields with cross pollination.
- ☆ Select pollinizers having capability to produce many flowers with abundant pollen grains (cv. Pendolino).
- ☆ Select resistant cultivars against frost.
- ☆ Plant those cultivars which are recommended, selected and authorized by the research institution for a specific and protected geographic area.
- ☆ Select drought resistant cultivars for rain-fed conditions or rocky soil and other cultivars for clay soil or irrigated ones.
- ☆ Select cultivars having different ripening period to facilitate. The harvesting operations and to allow processing of a special quality olive oil (olive oil "of a single cultivar or special combinations sometimes is preferable).

References

Caballero, J.M. and Del Río, C. 2008. The olive world germplasm bank of Spain. *Acta Hort.* (ISHS) **791**: 31-38.

FAO.2010. The second report on the state of the world's plant genetic resources for food and agriculture, Rome, Italy.

Grati-Kamoun, N., Mahmoud, F.L., Rebai, A., Gargouri, A., Panaud, O., Saar, A. 2006. Genetic diversity of Tunisian olive tree (*Olea europaea* L.) cultivars assessed by AFLP markers. *Genet.Resour. Crop Ev.* **10:** 265-275.

NBPGR, 2013. NBPGAR Annual Report, New Delhi, India.

Verma, M. K. 2006. Olive cultivation- deputation report submitted to ICAR (Personal information).

11

Orchard Establishment and Layout

Layout of the orchards is very important operation before establishing the commercial orcahrding. Since it is most permanent decision (along with varietal selection) in process of any orchard establishment. The trees will stay in the aspect and spacing as selected for them for many years to come. There are a number of factors which determine the physical layout of the orchard. The planting scheme is mainly depend on the cultivation system that will be applied to the orchard (high, medium or low density). For high density cultivation fertile soils and sufficient irrigation or rainfall is prerequisite, while these are not prerequisite for non-intensive cultivation systems. Traditionally olive trees have been planted at distances of approximately 10 m x 10 m or even larger which accommodates a limited number of trees per hectare (approx. 100 trees/ha). This high spacing may be suites to the areas which are characterized by a long period of drought so that high spacing may offer the trees the possibility to exploit all the available water. In present context this situation exists rarely therefore low spacing or high density plantation is best options to exploit all the natural resources judiciously. In general two are the main planting layouts: The existing traditional plantation of olive has a spacing of 7 x 7m, 6 x 8m, 8 x 8m or even 10 x 10 m (depending on the area) in contrast the high density orchards to be laid out at 5 x 6 m or 6 x 6m.

Orientation

While planning the aspect of the orchard the fact must keep in the mind that light and air penetration will be sufficient to increase the tree's health and crops. To achieve maximum light penetration into each row and each tree, the rows should be planted as close to a North-South direction as possible. That is, if driving a tractor up or down a row these would be facing either north or South which allows the light of the sun to penetrate the trees as it moves across the sky. The aspects depend on slopes or contours, but as a rule, try to keep as close as possible to North-South.

Location and Site

Proper selection of site is important. Selection may be made based on the following criteria.

- ☆ The location should be in a well established fruit growing region because one could get the benefit of experience of other growers and also get the benefit of selling the produce through co-operative organizations with other fruit growers.
- ☆ There should be processing and marketing facility near to orchard area.
- ☆ The climate should be suitable to grow commercial crops.
- ☆ Adequate water supply should be available round the year.

Before a grower selects a site for establishing a new orchard, the following factors must be assessed:

- ☆ Suitability of soil, its fertility, the nature of subsoil and soil depth.
- ☆ Site must have proper drainage and no water stagnation during rainy season
- ☆ Irrigation water must be of good quality.
- ☆ Whether the climatic conditions are suitable for the fruits to be grown and are whether site is free from the limiting factors such as cyclones, frost, hailstorms and strong hot winds.
- ☆ Whether there is assured demand in the market for the produce or product to be grown.
- ☆ Whether his orchard is a new venture or whether there are already other growers besides availability of labour.

Preliminary Operations

After selecting the suitable location and site the following operation to be done:

- ☆ Trees should be felled without leaving stumps or roots.
- ☆ The shrubs and other weedy growth to be cleared.
- ☆ Deep ploughing is essential to remove big roots.
- ☆ The lands should be thoroughly ploughed, leveled and manured. Leveling is important for economy of irrigation and preventing soil wash.
- ☆ In hills, the land should be divided into terraces depending upon the topography of the land and the leveling is done within the terraces. Terracing protects the land from erosion.
- ☆ If the soil is poor, it would be advisable to grow a green manure crop and plough it *in situ* so as to improve its physical and chemical conditions before planting operations are taken up.

Planning of an Orchard

A careful plan of the orchard is necessary for the most efficient and economic management. The following points should be borne in mind in preparing the plan.

- Optimum spacing to accommodate maximum number of trees per unit area. Stores and office building in the orchard should be constructed at the centre for proper supervision.
- There should be proper provision of pollinizer varieties in optimum ration.
- The irrigation channels must be laid out according to the slope of gradient for maximum use of water. For every 30m length of channel, 7.5 cm slope should be considered.
- Roads should occupy as low as space for transport activities. It is advisable to clear the space between wind break and first row of trees for the road.
- Short height trees should be placed at the front and tall at the back for easy visualization and to enhance the appearance of the orchard.
- To protect the orchards from abiotic stress like wind etc and animals a good fence is must. If possible choose live fences since these are economical then other kind of fences. The plants suitable for live fencing must have drought tolerance, disease resistant, easy to propagate from seed, quick growing, have dense foliage, able to sustain in severe pruning and should be thorny. *Agave, Prosopis juliflora, Pithecolobium dulce* and *Thevetia* if closely planted in 3 rows would serve as a good live fencing.
- To resist velocity of wind which causes severe ill-effects the rows of tall trees should be planted close together around the orchard to avoid moisture evaporation from the soil. The windbreak shows their maximum effectiveness for a distance about 4 times as great as its height but has some effect over 2 about that distance.
- To place windbreak appropriately the most effective plan would be that there should be windbreak in a double row of tall trees and place in alternately. The utmost care should be given in placement and must keep at least space as much as between the windbreak and the first row of the fruit trees as between fruit trees. To avoid competition between the wind breaks and fruit trees for moisture and nutrition the practice of digging a trench of 90 cm deep at a distance of 3m from the windbreak trees and prune and cut all the roots exposed and again fill up the trenches has to repeated for every 3 or 4 years.
- The windbreak tree must have a tendency grow as erect, tall and quick growing, hardy, and drought tolerant, mechanically strong and dense to offer maximum resistance to wind. Example: *Casuarina equisetifolia, Pterospermum acerifolium, Polyalthia longifolia, Eucalyptus globulus, Grevillea robusta, Azadirachta indica etc.*

The following steps to be followed for planting a healthy, fast growing olive orchard:

1. Placed FYM at each tree site at 2.5 to 3 cubic ft. (9 trees per cubic yard). Try to use well rotten FYM and avoid too fresh. Uniformly spread the manure over an area of 9 ft X 9 ft.
2. To enrich in mineral contents placed trace mineral such as Azomite (sometime called rock dust) at each tree site (contains excellent minerals) at up to 5 lbs per tree. This chemical is not water soluble and is naturally available to the tree roots as required. Agin this also has to be spread uniformely over the 9 ft x 9 ft area.
3. Add the required amount lime to the manure and trace minerals if soil is acidic in nature to bring its pH level to 7.0-8.0 (alkaline).
4. Deep rip has to be done for several times along the full length of the planting row to a depth of 2 ft or more and a width of at least 10 ft. The nutrients will be suitably mixed in as they drop down the ripper grooves. There should be end up with ten to twelve deep rips spread across the 10 ft width of the row. This preparation will give the roots an excellent start and fast growth.
5. Placed the tree at the same depth as it was in the pot and handle the plant with care so that root s will not damaged.
6. The soil should be press down firmly around the tree roots and make a depression to act as a watering basin.
7. The mulch such as coarse straw helps in conserving the water, cool the soil, and reduce weed growth. Beside this the mulches also contains plenty of nitrogen and other nutrients to feed the tree. These include lucerne, soya bean and pea hay. Keep the mulch 4″-6″ away from the base of the trunk to allow the tree to breathe. The soil microbes and earth worm decomposes the mulches over a period of time and then the nutrients are transferred into the soil. If planting to be done in an area with relatively long, cold, wet winters and short warm summers, only mulch very lightly or not at all. Too much mulch will conserve too much water.
8. There should be regular irrigation as per irrigation schedule but utmost care should be given to avid water logging since as excess water is the olive tree's worst enemy.

Laying Out of Orchards

The aim of best method of layout is to provide maximum number of trees per hectare, sufficient space for optimum growth and development of the trees and provide ease in cultural operations. The layout system can be categorized in to two broad categories *viz.* (a) vertical row planting pattern and (b) alternate row planting pattern. In the vertical row planting pattern (*e.g.* square system, rectangular system), the trees set in a row is exactly perpendicular to those trees set in their adjacent rows. In alternate row planting pattern (*i.e.* Hexagonal, Quincunx and Triangular), the trees in the adjacent rows are not exactly vertical instead the trees in the even rows are midway between those in the odd rows (Kumar, 1997).

The various layout systems used are the following:

a) Vertical Row Planting Pattern

1. Square System

This is most comely used planting system and is very easy to layout. Under this system trees are placed on each comer of a square irrespective of planting distance. The plants are planted exactly at right angle at each corner. The open central space can be utilized for raising the filler trees. This system permits inter cropping and cultivation in two directions. The major disadvantage of this system is that a lot of space is wasted in between the squares.

2. Rectangular System

In this system, trees are placed on each corner of a rectangle. In this system distance between any two rows is more than the distance between any two trees in a row, as in square system there is no equal distribution of space per tree. The wider alley spaces available between rows of trees permit easy intercultural operations and even the use of mechanical operations. The major disadvantage in this system is that two way inter cultivation is not possible.

b) Alternate Row Planting Pattern

3. Hexagonal System

This system is characterized by trees are placed in each comer of an equilateral triangle. This system differs from the square system in which the distance between the rows is less than the distance between the trees in the row, but distances from tree to tree in six directions remains the same. In all, 7 trees per hexagon are planted. Therefore this system is also called as 'septule' as a seventh tree is accommodated in the centre of hexagon. This system permits cultivation in three directions. The plants occupy the land fully without any waste as in square system. Beside this system allows planting of 15 per cent more trees than can be planted by square system in a given area. But it is difficult to layout by this method and also the watch and ward becomes difficult as one cannot see in all the directions from a point.

4. Diagonal or Quincunx System

This system has its unique attributes as one more plant in the centre of the square as compared to square system resulted able to accommodate double the number of plants, but does not provide equal spacing. The central (filler) tree chosen may be a short lived one. This system can be followed when the distance between the permanent trees is more than 10m. As there will be competition between permanent and filler trees, the filler trees should be removed after a few years when main trees come to bearing. The plant population is about double than the square system but there is difficulty in intercultural operations on account of the filler tree.

5. Triangular System

The trees are placed as in square system but the difference being that those in the even numbered rows are midway between those in the odd rows instead of opposite to them. Triangular system is based on the principle of isolateral triangle. The distance between any two adjacent trees in a row is equal to the perpendicular distance between any two adjacent rows. However, the vertical distance, between immediate two trees in the adjacent rows, is equal to the product of (1.118 x distance between two trees in a row). When compared to square system, each tree occupies more area and hence it accommodates few trees per hectare than the square system (Kumar, 1997).

6. Contour System

It is generally followed on the hills where the plants are planted along the contour across the slope. It particularly suits to land with undulated topography, where there is greater danger of erosion and irrigation of the orchard is difficult. The main purpose of this system is to minimize land erosion and to conserve soil moisture so as to make the slope fit for growing fruits and plantation crops. The contour line is so designed and graded in such a way that the flow of water in the irrigation channel becomes slow and thus finds time to penetrate into the, soil without causing erosion. Terrace system on the other hand refers to planting in flat strip of land formed across a sloping side of a hill, lying level along the contours. Terraced fields rise in steps one above the other and help to bring more area into productive use and also to prevent soil erosion. The width of the contour terrace varies according to the nature of the slope. If the slope becomes stiff, the width of terrace is narrower and vice-versa. The planting distance under the contour system may not be uniform.

During the last decade a new planting system is being tested in olive culture. It is called high density orchard system and it uses almost 1500-1800 trees per hectare. The main objective of this system is to reduce the cost of harvest, which in the traditional olive cultivation participate at around 40-50 per cent to the total cost of olive culture.

7. Hedge System

The layout is exactly same as rectangular system except that very wider spacing is maintained between rows and a very narrow spacing is followed between plants. This system permits easy movement men, material and machinery and also effective cultural operations due to wider spacing. Therefore, this system is especially suitable where machines are employed for various farm operations.

Calculation of number of trees required per unit area in different systems of layout (Naik, 2014).

The number of plants that can be accommodated by each of the systems in a unit area calculated by the formula shown against each system as under:

1. Square layout = S/L

   S= Unit surface (Area) L= Side of the square pattern (spacing)

2. Rectangular layout = $\frac{S}{(L1xL2)}$

   S = Unit surface (Area)

   L1 = Shorter side of the rectangle (spacing)

   L2 = Longer side of the rectangle (spacing)

3. Quincunx layout =

   $$\frac{\text{Area}}{\text{Spacing (Main trees)}} + \frac{\text{Area}}{\text{Spacing (Filler trees)}}$$

   Area of filler trees – (LXS) (BXS) L= Length of the orchard B= Breadth of the orchard S= Spacing between filler trees

4. Hexagonal system = $\frac{\text{Area x 115}}{\text{Spacing 100}}$

5. Triangular system = S/D 2 x0.8666

   S = Unit surface

   D = Length of the triangle side

6. Hedge system = Nx 100,000,000/Y(X+Z

   N = Number of lines in a hedge (double hedge – 2)

   Y = Distance between bushes in a line

   X = Distance between hedges

   Z = Distance between lines

References

Naik, B.H. 2014. Practical manual on fundamentals of horticulture and production technology of fruit crops. University ofAgricultural and Horticultural Sciences, Shimoga, pp. 6-9.

Kumar, N.1997. Introduction to horticulture. Rajalakshmi Publications, 28/5 – 693, Vepamoodu Junction, Nagercoil, pp. 15.47-15.50.

12
Production Systems

Traditional olive orchards in planted at a density of 70–80 trees/ha, with 3 trunks per tree and a medium–low productivity (Navarro and Parra, 2004). Harvesting in the traditional production system is very inefficient. The trees are almost always harvested by hand or by beating the fruit off with long poles onto nets. Because the trunks are so large, it is difficult and expensive to harvest the trees with mechanical shakers, because individual branches are shaken. Since the 1970s, new planting systems with 200–400 trees/ha and single-trunk trees came in to picture. By the end of the 20th Century, a new concept of olive orchard arose, based on a density of around 2000 trees/ha (Pastor *et al.* 2005; Tous *et al.* 2006). In these high-density orchards, trees are trained as a monocone with distances between trees lower than 2m; so 2–3 years after planting, they formed hedgerows. Vineyard straddle-harvesting machines are used to collect the fruit.

Advantages of High Density

1. The tree starts early production and initial harvest is starts from second year and based on soil and climatic condition the third and fourth year crops are heavy with full cropping in the fifth year.
2. Mechanical harvesting is possible which is more efficient and less costly in nature. Such as use of over-the row grape harvesters as well as newly developed harvesters in olive orchards leads faster and efficient harvests which almost eliminate expensive hand labour.
3. Higher yields and quicker payback – more trees per acre mean more production per acre. Yields can reach 4-6 tons per acre in mature orchards. Olive oil yields of 40 gallons per ton of olives can be expected.
4. Well suited to existing planting layouts.

5. Quality olive varieties – The specially selected IRTA® Arbequina, Arbosana and Koroneiki varieties not only have the compact growth habit for high density they have the quality of oil that makes them a premier extra virgin olive oil product.

High-density Production System

The high-density production system started around the early 1980s to reach full production faster in irrigated orchards. Hedgerows and a central-leader training system are being used, but most high-density orchards are still trained to the open-center form. Many orchards have been planted at different spacing combinations, usually with the row spacing closer than the between-row spacing to create a hedgerow with tree densities of 100 to 340 trees/acre (250– 840 trees/ha). The overall benefits have been significantly increases in yields per acre, usually double to triple what had been achieved previously. The orchards also come into full production sooner (>7–10 years), suffer less alternate bearing problems, and are more efficient to harvest. In high-density plantings, the fruit is primarily harvested with trunk shakers; however, there are some growers using other types of very large, single-sided or overthrow comb-type harvesters. The high density system has wide adaptability and will accommodate any olive variety, soil type, terrain, or training system. On steep terrain where tractors and mechanical shakers cannot function, the trees can be harvested by hand or with various assisted hand-harvesting methods. The main problems with the high density system are that the trees still require 8 to 10 years for full production. Successful super high-density olive orchards were pioneered by the Agromillora nursery and growers in Catalonia, Spain. This system uses specific varieties planted at tree spacing's of 3 to 5 feet (0.9–1.5 m) within the row to 12 to 13 feet (3.7–3.9 m) between rows for 670 to 1210 trees/acre (1655–2990 trees/ha) (Vossen, 2007).

The three best known varieties for super high density systems, observed to date are 'Arbequina', 'Arbosana', and 'Koroneiki'.

Characteristics of these Varieties

- ☆ They can be grown in an upright fashion,
- ☆ Able to train as a central leader,
- ☆ Throw few large lateral branches and yet are compact compared with other varieties,
- ☆ Precocious, tend to produce a good crop every year, start bearing at an early age, and have excellent oil quality characteristics.

However, these are not dwarf varieties and, if grown in a wider spacing with different management, they will eventually become just as large as any other olive trees (Vossen, 2002). The 'Arbequina' variety has been the most widely planted variety for several years in both high-density and especially in the super high-density systems. The 'Arbosana' variety has fruit that look very much like 'Arbequina', but matures 3 weeks later and has 25 per cent less vigour than 'Arbequina'. 'Koroneiki' is the primary oil variety of Greece, produces excellent oil, and has annual heavy cropping and precocious bearing. It has about the same vigour as 'Arbequina', but smaller fruit size, and requires greater force for fruit removal. Super high-density orchards are coming into bearing in the second year, with full production in the fourth or fifth year. Their yields ranged from 1.2 to 7.8 t/acre (2.7– 17.5 t/ha). It is very likely that production levels of about 4 t/acre (9 t/ha) can be maintained with this system. However, the long-term production of the super high density orchard system is not known. The main problem is maintaining light exposure into the trees' interior canopies. This system has been quite successful the first 14 years, but plant manipulation techniques to maintain high production have a steep learning curve. The super high-density system requires fairly flat ground and requires the ability to control tree vigour through pruning, fertility management, and controlled deficit irrigation. There is a higher capital investment needed because of the extra trees, trellises, and more closely spaced irrigation system. The super high-density system also requires a high degree of technical skill by the farm manager because of the requirement to manage the size, light exposure, and disease susceptibility of the closely spaced trees (Vossen *et al.*, 2004).

The super high- density system is not simple; it requires a great deal of skill, and more investment capital, but with the following main points if correctly followed, it can be successful:

1. **The site***:* It should well-drained soil that is not excessively steep in order to accommodate the mechanical harvester. Excessively fertile "bottom" ground that is very deep will likely not limit the vigour of the olives and could lead to poor fruiting and excessive shading.
2. **Planting**: Olives are semi-deciduous so they can be planted during most months of the year. The plants are produced in small pots for easy transplanting to the field. Once the ground is ready and the irrigation system is installed olives can be planted.
3. **The varieties***:* Select the appropriate varieties such as Arbequina, Arbosana, and Koroneiki depending upon climatic condition and availability of genotypes.
4. **Rootstocks:** 'Arbosana' and 'Limoncillo' rootstocks have an ability to reduce the vigour of 'Arbequina-i·18®' cultivar because they promote higher productivity (kg/cm^2) for this clone. These rootstocks are a potential choice for establishing a super high density orchard.
5. **Tree spacing***:* It is depends on variety and its characteristics along with climatic influences. If the soil is deep and fertile, the trees should be planted at a slightly wider spacing. Minimum spacing = 3 ft. x 12 ft. - Maximum spacing = 4 ft. x 14 ft.
6. **Push the young trees***:* By maintaining adequate moisture and good weed control leading the trees to grow rapidly and fill their allotted space within the first 3-4 years.
7. **Pruning**: Varieties used in high-density olive orchards are pruned to a central leader. A bamboo, wooden stake or metal pole must support the

plant because of the low vigour and high productivity. The tree stakes are to be supported by a one-wire trellis system with end poles and metal stakes spaced for support of the wire down the row. The trees are not headed when planted and are allowed to grow up the stakes in a central leader fashion. The tree canopy is kept to about two feet above the orchard floor and topped to a level of 7 feet.

8. **Training**: Tress should be trained to an upright central leader form that keeps the lower 3 feet clean of all lateral branches for good closure of the mechanical harvester catch frame. Very little pruning is done in the first 3 years; the central leader is simply tied to a support as it grows vertically. No large lateral branches are ever allowed to grow. Trees are mechanically topped in the summer to maintain a maximum height for the straddle harvester.
9. **Fertility levels**: The fertility level should be maintaining optimally especially of nitrogen which are reduced after the 3rd year in order to create less vegetative vigour.
10. **Controlled deficit irrigation:** Management practices are followed after the 4th year to limit vigour, save water, and maximize oil quality.
11. **Disease monitoring:** As olive is attacked by many diseases, utmost care should be given to check theses disease and especially peacock spot should never be allowed to develop.

References

Navarro, C. and Parra, M.A. 2008. Plantation. p. 189-238. In D. Barranco, R. Fernández-Escobar and L. Rallo (eds.). El Cultivo del Olivo (6th ed.). Mundi-Prensa and Junta de Andalucía. Madrid.

Pastor, M (ed.). 2005. Cultivo del olivo con riego localizado. Junta de Andalucía and Mundi- Prensa. Madrid.

Tous, J., Romero, A. and Hermoso, J.F. 2006. The hedgerow system for olive growing. *Olea* **26**: 20-26.

Vasan, P. 2007. Olive oil: history, production, and characteristics of the world's classic oils. *HortScience* **42**: (5) 1093-1100.

Vossen, P. M. 2002. Super-high-density olive oil production. OLINT Magazine Special English Edition 1(Oct.).

Vossen, P. M., J. H. Connell, K. Klonsky, and P. Livingston. 2004. Sample costs to establish a super high-density olive orchard and produce oil, Sacramento Valley. University of California, Davis, Department of Agricultural and Resource Economics.

13
Propagation

Plant propagation is the process of multiplication and production of the plants. The fruit plants are propagated both by sexual and a sexual methods. Sexual methds described as the union of the pollen and egg, drawing from the genes of two parents to create a new, third individual. It involves the reproductive parts such as floral parts of a plant. Asexual method also known as vegetative propagation which results genetically identical plants as its parent. It involves the vegetative parts of a plant as leaves, stems or roots. The major advantages of sexual propagation are that it is cheaper and quicker and only way to obtain new varieties and hybrid vigor. In some of the fruit crops this is only viable method for propagation beside this also a way to avoid transmission of certain diseases (viruses *etc.*). Vegetative propagation is easier and faster in most of the fruit crops and it bypasses the juvenile characteristics. Olive trees are propagated by several methods mentions below:

Sexual Method

Seeds

The partially ripe fruits are collected during September-October. The stones are separated from their pulp by treatement of caustic soda containing 10 per cent Na OH or KOH in a dilute HCI. The stones are thoroughly washed in running water to remove all traces of chemicals. Then they are immediately sown in raised nursery beds at a spacing of 15 cm from row-to –row and 5cm from seed-to –seed. The nursery beds are mulched and regularly irrigated. The seeds germinate with the onset of spring season. However, a few seeds germinate 1-2 years after sowing.

At CITH, Srinagar, an experiment was conducted to see the effect of different scarification treatments on olive seed germination and found GA_3 500 ppm for 12 h dip best for Coratina and Pendalino cultivars (Lal *et al.*, 2015).

Asexual Method

Stem Cuttings

It is the simplest method of vegetative propagation and commonly practised. The 12- 14 inch long and 1-3 inch wide cuttings from the two year old wood of a mature tree is treated with a root promoting hormone, planted in a light rooting medium and kept moist. Trees grown from such cuttings can be further grafted with wood from other genotypes. Cutting grown trees bear fruit in about four years.

Steps in Producing Stem Cuttings

- ✰ Take cuttings from well developed and vigorous disease free mother plant.
- ✰ Take cuttings from the middle or bottom parts of the tree.
- ✰ Avoid taking cuttings from the fruited buds. Vegetative buds are highly capable of forming roots.
- ✰ Choose a shoot of 6 to 12 month old, 45 to 60 cm. long and 3 to 8 mm in diameter.
- ✰ Remove the tender portion from the top and hard portion from the base of the shoot.
- ✰ Make 3 cuttings (basal, medial, and apical) from the shoot, each measuring 10 to 15 cm. long. The apical (terminal) cutting results better in Spring while the basal and the medial cuttings perform better in Summer and Autumn, due to presence of more carbohydrate and hormones on them.
- ✰ Give a slant cut on the top and straight cut on the bottom of cuttings to mark the correct polarity. The bottom cut is made just below a node, otherwise it will not root.
- ✰ Remove all leaves except 2 to 3 pairs on the top portion of the cuttings. The cutting should be fresh (not more than two days from the day of detachment) and be kept in the shade and moist condition.
- ✰ Do not take cuttings during very hot and very cold seasons.
- ✰ Cuttings can be put in the benches or beds 2 to 3 times each in Spring and Autumn.

Hartmann (1946) developed a technique for olive tree propagation by leafy-stem cuttings. This technique uses 1-year-old stems, which have a lower natural rooting capacity than hardwood cuttings, and obtains many olive plants from the same tree. This method is based on the application of root-promoting products, which stimulate adventitious root formation in small cuttings with leaves retained at the upper end. The cuttings are then placed in a mist bed, which keeps a thin film of water on the leaves, thereby reducing leaf temperature through evaporative cooling and transpiration and avoiding leaf dehydration. Root formation is stimulated by the application of bottom heat (23 to 25 °C) and coarse mineral components (perlite). Rooting success mainly depends on two factors: cultivar and hormone–nutrition balance (Hartmann, 1946; Hartmann *et al.*, 2002; Khabou and Trigui, 1999).

Mist Propagation Unit

Propagation by cuttings requires a special environment where these cuttings are induced to root. With this process, the parenchymatic cells become meristematic and division of cells starts. The indigenous auxins and other components available in the cutting stimulate root formation. This stimulation is possible only when determinate environment parameters are satisfied as temperature around 20° to 25° C, high relative humidity 80° to 100°, and rooting 'media' with good water holding capacity, porosity, good drainage (needs to keep for long time),and adequate density. To satisfy these parameters it is necessary to have a bench with bottom heating system, mist spray system, and cooling system. All these systems including the rooting bench, where cuttings are normally placed, are called 'Mist Propagation Unit' (MPU). In other olive growing areas there are glasshouses or plastic houses which are well equipped and authomatized; these structures are able to produce more rooted cuttings at one time. To obtain good results with small MPU in different climatic conditions where infrastructures are lacking it however is not so easy. There are four determinate periods of climate such as hot and dry, hot and wet, cold and wet, and cold and dry. Moreover the local nurseries have two main sources available: water from channels and electricity. Other facilities are not suitable for this kind of activity (Solar energy, gas cylinders, *etc.*). But even these two sources can be used to some extent (the water from channel is having multipurpose functions; electricity is for houses or for small electric engines.). In this condition it is necessary to find a MPU that should be simple, cheap, and able to give good results.

Hardwood Cuttings

Hardwood cuttings can be made from 2 or 3 year old wood about an inch in diameter, and 8 to 12 inches long just prior to warm spring weather. All leaves should be removed. Soaking the ends of the cuttings in hormone solutions, followed by storage in moist sawdust or sand at low temperature for a month to help induce callus formation is often used. Cuttings are then placed in the nursery to root, often being lined out in well worked, friable soil. The cuttings should be mostly buried, and kept moist but not wet. Rooting will occur over several months. Trees are then dug bare root and containerized or planted into the orchard.

Rooted Truncheons

The truncheon system is also used as a low-tech system for olive tree propagation. Limbs 3 or 4 inches in diameter are removed from trees and cut into 12 inch pieces, and then planted horizontally in soft, well tilled friable soil. Usually several shoots with an accompanying root system will grow. They can be separated, and grown for another year before being planted in the orchard.

Rooted Ovule

Swellings found on the trunk of the olive tree, known as "ovuli", can be cut off and planted in December-January. These structures contain both adventitious root initials and dormant buds so that new root and shoot systems can develop. This practice is damaging to the parent tree, and is not used very often.

Suckers

Suckers with a small piece of root can be removed from the trunk of the tree in the winter and grown in the nursery for a year before planting into the orchard. Suckers are shoots that are originated on ovules or basal parts of trunks of the older trees. Instead of removing the ovules, they can be induced to produce rooted suckers, even when they remain *in situ*. The rooted suckers can be produced by another way also. The base of tree is covered with a light layer of soil, or strongly binding the base of the tree with one or two turns of wire, or ringing the bark of the trunk. The suckers removed from the mother tree can be taken to the nursery for further development, or they can be planted directly in their permanent site. This method of propagation is not a healthy practice as it weakens the mother tree and plants produced in this way show juvenile characteristics.

Root Piece

Root pieces of about 5 to 10 cm in diameter, each weighing 2 to 5 kg are put into the soil consisting well fermented manure at the onset of rainy season. Small shoots emerge from those pieces of the roots.

Root Cuttings

The roots weighing 200 to 400 g are put in the plastic bags. These root cuttings are developed as planting materials for the supply to another area. This is modified method of the root piece technique.

Branch Piece

A piece of branch measuring 5 to 7 cm in diameter and 30 to 40 cm in length is taken. The pit is dug and riled it after one month with a mixture of soil, and well rotten manure or cow dung. The two-third parts of the branch are buried below the soil level at the permanent site and one-third of the top portion is exposed above the soil. This method is also applied during winter and sprouts during April-May. The shoots emerged are kept as such for 2 years and then, only one shoot is selected to grow by removing all other shoots.

Grafting/Budding

This technique is used in olive varieties with a low rooting percentage in cuttings. However it requires more space and longer time periods for plant production. The main rootstock sources are seedlings of wild or cultivated olive varieties. The worker must be skilled in grafting; therefore training and subsequent practice is very important. Correct and efficient equipment and materials are required, appropriate to the type of grafting to be carried out (knife, tying materials etc). The rootstock and scion should be true-to-type and compatible. The correct time of the year is important (when the cambium of the rootstock must be active). Scion/rootstock characteristics and guidelines for successful grafting/budding:

- ✰ The rootstock should be in its active growth stage *i.e.* time of budding and grafting is important. Best periods for budding (May-June) and grafting (Nov.-Dec.). The rootstock should be 9 to 18 months old.

- ☆ The appropriate time to collect the scion-wood is at the active growth stage of the mother plant and rootstock (seedling) must be healthy (free from diseases and insects), vigorous, with straight trunk, have smooth texture and clean bark, thickness of 0.6-1cm diameter with a well developed root system.
- ☆ To avoid drying, the collected graft sticks should be wrapped in a wet clean cloth and put in a polyethylene bag.
- ☆ The buds should not be removed from the bud stick prior to budding operation.
- ☆ The cut should be done first on the rootstock and then the bud from the stick. They should then be immediately placed inside the bark of the rootstock for better results.
- ☆ The final cut of the scion should be done according to the shape and size of the cut made in the rootstock.
- ☆ The point of budding/grafting must be at least 15 cm above the ground.
- ☆ The rootstock-scion union should be wrapped with grafting and budding tape in such a way that no air and water can enter into the wound part.
- ☆ Unwrapping of the wound should be done after the rootstock-scion union is completed, usually after three weeks.
- ☆ The upright growth of the scion is enhanced by removing lateral branches or the buds developed below the rootstock-scion union.

The after-care of grafts is also a significant factor of success and must be taken care as following:

- ☆ Correct environmental factors (temperature, humidity, shading, ventilation).
- ☆ Prevention of drying out of the cut surfaces (tying materials, tie-in properly).
- ☆ Prevention of pest and disease infection.
- ☆ The tying-in material must be removed in 3 to 4 weeks after grafting/budding otherwise it will bite into the plant tissue if retained too long and cause constriction as the stem girth increases.
- ☆ Optimum irrigation-fertilization (fertigation).
- ☆ Weed Control - use black polyethylene on the ground to avoid weeds.
- ☆ Caning and tying-in to give further support to the graft union and the scion growth in order to avoid bent stems.
- ☆ Shading for hardening-off (weaning-off) and proper training of the grafted/budded plants in the nursery facilities with a single trunk and 2 or 3 branches about 1 m above the ground.

Raising Seedlings as Rootstocks

The seedling rootstocks can be obtained from wild-olive stones or from stones of cultivated varieties. A greater percentage of germination and more vigorous seedlings are obtained from the former. The wild species, *Olea cuspidata* available in the North West Himalayan region has proven successful for using as rootstocks.

The followings are the steps for raising the seedlings:

- ☆ Collect ripe fruits from the tree for seeds during October to December under Indian conditions.
- ☆ Smash or crush the fruit gently to remove flesh (skin and pulp) within 6 hours after picking.
- ☆ Clean the seeds with sand and water to remove the sticky materials or with a solution of NaOH and water (250 g of NaOH for 100 kg seeds and 100 litres of water).
- ☆ Store the cleaned seed in a dry place in a layer 4 to 5 cm thickness, covered with paper sheets and free from insect.
- ☆ Soak the seeds in water for 15 to 20 days before sowing in the beds. Change waters two times every day while soaking; you may add in the water 5 ml NAA (Naphthalene acetic acid) in 10 litters of water for 10 kg seeds to stimulate germination.
- ☆ The pits can also be scarified (cracked or clipped) and stratified (soaked in water and put in most sand) for quicker germination.

Commonly used root stocks:

1. *Olea europaea* var. *Oleaster* seedlings (wild form)
2. *Olea cuspidata* (Indian native olive)

Patch Budding for Top Working Olives

Probably the simplest to perform among the various methods of budding due to ease in removing or preparing rectangular patches of bark. It is widely used in plants with thick bark that can be easily separated from the wood. The method involves the complete removal of a rectangle-shaped patch of bark with the longer sides parallel to the axis of the stem of the rootstock. It is then replaced with a bud patch of the same size from a bud stick. The patch of bud is cut from both the rootstock and the bud stick by two parallel horizontal cuts either with one stroke of a double-bladed knife or two strokes when using a single-bladed knife. With vertical stroke of a knife, both horizontal cuts are connected at each side. The bud patch is carefully removed intact and inserted into the rootstock. Patch budding can be used form top working olives alone or in conjunction with bark grafting. It must be done when the bark slips or separates easily from the wood. It is slower than "T" budding and chip budding, but it is widely used for thick-barked species such as the walnut. Results are best when done in late fall or early spring. Patch budding can be more desirable than bark grafting because.

- ☆ Less scion material is required
- ☆ It is faster and less expensive
- ☆ It can be done during a longer time period

- No tree seal required
- Damage from wood boring insect reduced

The disadvantage of patch budding is that when it is done on larger limbs, vigorous new growth from patched buds may be more likely to break off.

Micro-propagation

In vitro propagation allows producing high quality and rapidly growing plants. The first study on olive micro propagation was reported by Rugini (1984). The development of specific olive medium for axillary bud stimulation and subsequent shoot multiplication marked an important step forward in the improvement of olive micro propagation. Over the past decade, many advances have been made for micro propagation from mature olive trees. Micro propagation requires the availability of well equipped laboratory where explants are prepared in media with a combination of chemical nutrients, plant hormones and other compounds for their multiplications at specific light intensity and temperature. Micro propagation of olive is made in 4 phases:

- Preparation of explants (uninodal explants),
- Proliferation of the explants *in 'vitro'*,
- Rooting of explants *in 'vitro'*, and
- Acclimatization. *In vitro* propagation allows producing high quality and rapidly growing plants.

The micro propagation techniques are preferred over the conventional asexual propagation methods because of the following reasons:

- In the micro propagation method, only a small amount of tissue is required to regenerate millions of clonal plants in a year.
- Micro propagation is also used as a method to develop resistance in many species.
- *In vitro* stock can be quickly proliferated as it is season independent.
- Long term storage of valuable germplasm possible.

The factors that affect micro propagation are:

- Genotype and the physiological status of the plant *e.g.* plants with vigorous germination are more suitable for micro-propagation.

Criteria for Selecting Propagation Material

- It is essential to study the morphological, physiological and agronomic characteristics of the strain or clone that it is desired to propagate.
- Only moderately vigorous branches must be selected for scions and cuttings. They must not have juvenile characteristics that prolong the non-producing phase of the future plants.

☆ The selected material should be absolutely healthy, *i.e.* free from insects and diseases. It is desirable to set up plots of selected mother trees close to each nursery and attention must be given to its biological measures and sanitary controls. Where there is no mist propagation facility, the desired plants can be produced by adopting other methods such as budding, grafting and planting stem or roots in the nursery beds or directly at the permanent site.

Mother Plant Plot

To obtained continuous quality scion wood there must be a mother plant block. The rapid development of mother plants is necessary to obtain maximum number of cuttings. The mother tree can be maintained as bushy from the ground level and to develop shoots a bit above on the single stem. However, it is advisable to adopt maximum planting density per unit area compatible with cultivation practices and with sufficient light reaching the top of the mother plants. The mother plants should be well identified and sure of the variety. The mother plant plot should be situated near the propagation unit and should have irrigation facility. The site for mother plant plot should not be located at extreme hot or cold place. The soil should be friable and tillaged according to the density of water.

Establishment of Mother Plant Block

Plough the land and level the soil. Dig 45 to 60 cm deep and wide pits at 2x2m or 2x3m or 2x4m distance depending upon the planned methods of cultivation and training system of the plants. Mix the top soil with 7 kg of FYM, 200 g supper phosphate, 200 g potassium sulphate for one pit. Put normal soil into the bottom part, mixture in the middle part and again normal soil on the top portion of the pit. Plant one vigorous, healthy and well identified sapling in a pit and press the soil around the plant to remove air. Irrigate plants immediately after planting and then 4 times at 15 days interval. Irrigation once in a month during dry season is necessary. Give staking to the plant to protect it from wind. Plough or hoe the soil 10 cm deep, 4 to 5 times in a year *i.e.* during November, April, May, June and September. The weeds around plants should be removed by hands. The weeds harbour insects, disease and compete with the plants for nutrition and moisture. Do not prune mother plants at all for two years, cut the stem at ground level in the third year. A number of shoots will arise at the ground level and the plant will develop a bushy form. Apply following fertilizers for 1000 sq. m area of the plants every year; half dose in Sept –Oct. and other half during March- April: 21 kg. urea (46 per cent N), 15 kg. ammonium nitrate (33 per cent N) in) 15 kg. Supper Phosphate (46 per cent P_2O_5), 15 kg. potassium sulphate (50 per cent 2O) add manure 3 cubic meter in every two years, and give foliar sprays 4 times in a year, 2 times in Autumn and 2 times in Spring to accelerate vegetative growth. This condition will give good shoots, good carbohydrate and root age percentage. The cutting can be taken all the year round except in very cold or very hot season. Cut whole scion shoots of appropriate size at surface level for putting in nursery leaving only thin branches for next season. The cuttings should be kept under shade (Fergusson 1994).

References

Fergusson, S.S. and Martin (ed,). 1994. Olive production manual, publication n.3353 of the University of California USA.

Hartmann, H.T. 1946. The use of root-promoting substances in the propagation of olives by soft-wood cuttings. *Proc. Amer. Soc. Hort. Sci.***48**: 303–308.

Hartmann, H.T., Kester, D.E., Davies, F.T., Geneve, R.L. 2002. Plant propagation. Principles and practices (Prentice Hall, Upper Saddle River, NJ).

Khabou, W., Trigui, A. 1999. Optimisation of the hardwood-cutting as a method of olive tree multiplication. *Acta Hort.* **474**: 55–58.

Rugini, E.1984. *In vitro* propagation of some olive cultivars with different root ability, and medium development using analytical data from shoot and embryos. *Sci. Hort.* **24**: 123-134.

14
Weed Management

Weed management in olive orchards is the crucial intercultural activity which reduces the negative impact of weeds on trees growth, prevent build-up of hard pan, and reduce plant debris at the base of the tree to facilitate harvest. Therefore, it is essential to manage the weeds in the olive orchards to improve the tree growth, yield and quality of fruits. However, weeds affect the crops in many ways, briefly as given below:

Harmful Impact of Weeds on Olive

Impact on Water and Nutrients Use Efficiency

Weeds compete directly or indirectly with the use of water and nutrients applied in the olive orchards. This occurs mainly during the period of water shortage and fruit development, in spring and summer, and is more intense where root density is greater. The effect on small olive plants is more as compared to mature trees. Therefore, it is desirable to undertake the cultural practices aimed at maintaining cover to prevent erosion and soil degradation and to encourage species diversity, adequate weed control is clearly a priority to avoid crop losses.

Impact on Harvesting and other Cultural Operations

Heavy weeds infestation on the ground makes harvesting very expensive to fallen olive fruits; it is also difficult to check drippers, to prune the trees or to apply plant protection products. These drawbacks are more evident underneath the canopies of the trees while they are almost insignificant in the orchard lanes. Consequently, closer control is required under the canopies than along the middle of the lanes. Some species can also cause physical discomfort to operators, for instance thorny species or species such as *Capnophyllum peregrinum* which cause skin rashes.

Impact on Pests and Diseases and of Climatic Damage

The presence of weed covers, leads to higher environmental humidity and a greater incidence of air-borne fungi like olive leaf spot. Higher incidence of olive *Psyllid* has also been observed associated with weed infested orchards. Mature and woody weed species harbours and protects more number of rodents like, rabbits, moles and mice in the orchards. The damage from spring frosts are greater in olive orchards covered with vegetation because the frost persists longer and more intense with vegetation.

Some Direct and Indirect Benefits Associated

- Formation of soil and reduced rate of erosion.
- Improves the biomass *i.e.*, fauna and biodiversity.
- Provides organic matter and fix nutrients and atmospheric CO_2, results in the reduction of green house gases produced by industrial and urban activity.

Weeds Management Strategies

Weed management in olive differs for each orchard and varied weed species, soil type, irrigation method, amount of control desired, terrain and the appearance of the orchard. For example, winter annuals are least troublesome because there is generally enough moisture during winter to support both the tree and the weed. They can be managed in the spring. Summer annuals, biennials, and perennials require stricter management. Perennials in particular should be eliminated completely.

Weed Management in Young Orchards

Weed management is critical around young trees where weeds compete for nutrients, water, and light. Weedy orchards may take 1 to 2 years longer than those that are weed-free to become economically productive. (From an economic standpoint, however, it is important to compare the costs of weed management with the benefits of earlier production.) To control weeds after trees are planted and before bearing, apply a pre-emergence herbicide to either a square or circle around each tree at least 4–6′ (1.2 to 1.8m) across, or as a band down the tree row. Selective post-emergence herbicides are available for the control of most annual and perennial grasses. Young trees need to be protected from contact by some post-emergence sprays. It is best to mow and/or cultivate (see below) in conjunction with the use of herbicides. Mow weeds when they get 6 to 8″ (15 to 20cm) high, usually about four to eight times a year. Cultivation is required when weed seeds germinate after irrigation.

Weed Management in Established Orchards

The established trees are more tolerant of many herbicides than newly planted trees, thus increasing the options available for weed control.

- ☆ Generally weeds are controlled between tree rows by discing or mowing, and a basal treatment of herbicide is applied around each tree or in a strip application down the tree row.
- ☆ Pre-emergence herbicides can be applied either alone, in combinations of herbicides in fall after harvest, split into two applications as described above, or in winter with a post-emergence herbicide.
- ☆ Delay the pre-emergence application in winter until most weeds have germinated, and then to add a post-emergence herbicide. This allows longer weed control into the summer yet does not allow much competition from weeds to the tree.
- ☆ For greatest safety, herbicide sprays should be directed only at the soil or at the weed foliage, not at the tree leaves or 1- to 2-year-old wood.
- ☆ In orchards where tree rows are mulched or sprayed, there are often few weeds to treat, and a sprayer with a weed detection system can be used to reduce herbicide use.
- ☆ For treatment of small areas, especially for perennial weeds, a backpack sprayer or low-volume controlled droplet applicator can be used.
- ☆ Extreme care needs to be exercised to avoid drift of herbicides to tree leaves or green stems.
- ☆ Frequent wetting of the soil promotes more rapid herbicide degradation in the soil. Herbicide degradation is generally faster in moist, warm soils than in dry, cold soils.
- ☆ Degradation is also more rapid under drip emitters or micro-sprinklers than under furrow or sprinkler irrigation.
- ☆ The first irrigation following an herbicide application is the most critical in terms of how far the pre-emergence herbicide is moved into the soil; subsequent irrigation is less important to the movement of the herbicide.
- ☆ The optimum amount of water for herbicide activity is from 0.5 to 1″ (13 to 25mm). Greater amounts of water (3 to 6″, 76 to 152mm) could move the herbicide far enough into the soil, especially in sandy areas, that it is absorbed by the tree's roots.

Soil Characteristics and Weed Management

Soil characteristics play important role in weed management. Soil texture as well as organic matter content influences the intensity of weed population, the number and timing of cultivations required, and the activity and residual effects of herbicides. The irrigation method, terrain, amount of water applied, and pattern of rainfall also affect the frequency and timing of cultivation as well as the selection of chemicals and their residual activities.

Tillage Practices for Weed Management

Tillage practices are used to weed management as well as improving soil aeration. Tillage practices are very effective for controlling against annual and biennial weeds, but less effective against perennials. It is highly useful for controlling flora adapted to no tillage, for example *Conyza* spp. of which populations resistant or tolerant to herbicides like simazine, diuron or glyphosate tend to appear in non-tilled plots with bare soil.

Mulching and Cover Crops as a Tools for Weed Management

Mulches and cover crops are the effective tools used in controlling weeds in perennial orchard crops. Mulch layer blocks light to reach the weeds, prevents germination or growth of the weed seeds. They create more uniform moisture conditions and conserve water, which in turn promotes tree growth. Soil temperature is better maintained and organic material is added to the soil on breakdown. However, mulches may also provide a good habitat for gophers, voles, field mice, and snakes or be a source of new weed seed that came with the mulch.

Different type of mulches vary in the blocking the light to the weeds. Organic mulches includes, straw, green waste, composted wood chips, sawdust, newspaper etc are effective when applied in a 4" (10cm) or more thick layer. Organic mulches are good in many ways like they degrade easily, they must be replenished annually, increases the cost of cultivation appreciably. As mulches degrade they become a perfect growth medium for weed species such as common groundsel, prickly lettuce, common sow thistle, and panicle-leaf willow herb. Always apply mulches when the soil surface is free of weeds. Synthetic mulches like, polyethylene, polypropylene, or polyester etc are very to use and more effective. These are commercially applied in the orchard crops. However, mulches do not control perennial weed growth unless they are completely away from light. Some woven fabric mulches also offer excellent weed management for several years, but the initial cost is very high.

Cover crops are used in orchards to replace the resident weed population. The type of cover crops or species differs from one orchard to another. Therefore, selection is crucial for effective management of weed population in olive orchards. Cover crops like wheat, oat, cereal rye, or barley, Blando brome grass, Zorro fescue, rose clover, and subterranean clovers do not compete with the trees. The cover crops are sown in late September through mid-November or resow if mowed in January or early February and then allowed to re-grow into April and May. Mowing after the seeds mature ensures seeds for the next season. Avoid invasive plants such as white clover and Bermuda grass in a ground cover. Sometimes larger-seeded cover crops such as bell bean, purple or common vetch, or crimson clover are planted in orchards and tilled in as green manure. Perennial grasses (tall fescue, Berber orchard grass, or perennial ryegrass) may also be grown but will require summer irrigation and may compete with tree growth. Keep cover crops away from the trees. Changing cover crop species reduces the potential for build-up of disease pathogens, weeds, rodents, and insect pests. Cover crops can be planted between the tree rows, and in the spring a "mow-and-throw" mower can be used to cut the cover crops and throw them in the tree rows. This works well if the mulch layer is thick.

Use of Herbicides for Weed Management

Herbicides are considered one of the most effective and widely used chemicals for weed management. Application of herbicides provide effective, economical and reliable control measures for most of the weed species including monocot, dicot, annual, biennial or perennial weed species. For providing effective control measures, correct identification of the weed species and its control through use of specific herbicide or broad spectrum herbicides is essential. Herbicides are mainly classified into two categories *i.e.* pre-emergence herbicides- those that are active against germinating weed seeds and post-emergence herbicides-those that are active on growing plants. Some herbicides have both pre- and post-emergence activity. These herbicides vary in their ability to control different weed species.

Pre-emergence Herbicides

Pre-emergence herbicides are applied to bare soil and are leached into the soil with rain or irrigation where they are active against germinating weed seeds. They must be applied by water (rainfall or irrigation) into the top 1 to 3″ (2.5 to 7.6 cm) of soil where weed seeds germinate. If herbicides remain on the soil surface without incorporation, some will degrade rapidly from sunlight. Weeds that emerge while the herbicide is on the surface, before it is activated by rain or irrigation, will not be controlled. Large weed seeds, such as wild oat, may germinate in the soil below the herbicide zone and still be able to emerge. Some pre-emergence herbicides must be incorporated within a week by water to be effective, while some can wait longer. Some cannot be incorporated mechanically without reducing their effectiveness. Pre-emergence herbicides can control weed emergence from several weeks up to a year, depending on annual rainfall, solubility of the material, soil properties, frequency and method of irrigation, weed species, and dosage applied. Prolonged moist conditions around low-volume emitters promote the breakdown and leaching of herbicides. Splitting a pre-emergence herbicide into two applications (with the same total dosage) can prolong control, particularly in areas with heavy rainfall, on sandy soils, in early fall treatments, or in orchards with a heavy growth of summer annuals. Split applications can be made by using one half to two thirds of the total required dosage in the fall and the remainder the following spring. Pre-emergence herbicides are more phytotoxic to plants in sandy soils and in soils low in organic matter than in soils high in clay and organic matter. They leach from the surface more rapidly in sandy soils than in clay soils, which may allow some weeds to germinate above the herbicide. In orchards with sandy soils, therefore, split treatments are safer for the trees and provide longer control. Because pre-emergence herbicides can persist in the soil for a few months to a year, their use should be discontinued 1 to 2 years before removing the orchard if the soil is to be replanted. Where a tree must be replaced, untreated soil should be backfilled around the root of the new tree to avoid damaging the tree itself.

Post-emergence Herbicides

Post-emergence herbicides are applied to control weeds already growing in the orchard. They may be contact herbicides or translocated (systemic) herbicides. Contact herbicides kill only the parts of the plants that are actually sprayed; good coverage and wetting are therefore essential. A single spray kills susceptible annuals; retreatment is necessary if regenerating perennials are present or if annuals re-establish themselves from seed. Contact herbicides are most effective on young weeds where it is easier to get good coverage and less material is needed. Tran located herbicides are transported from the sprayed part to the rest of the plant, including its roots, growing points, and storage structures. Thorough coverage is therefore not as important as with contact herbicides. Translocated herbicides are effective on both young and old weeds. No herbicide is effective on old, dusty mature weeds. Because different herbicides work in different ways and on different weeds, they are sometimes combined. No single herbicide works on all weeds. In many instances, combinations or sequential applications of different herbicides provide better control than one product alone.

Other Methods of Weed Management

Flame Weeding (Flaming)

Flaming is a method that can be used to control very young weed seedlings. Propane fuelled flamers are the most common. Heat causes the cell sap of plants to expand, rupturing the cell walls. This process happens in most plant tissue at 130°F (54.5°C). Flaming is not intended to burn the weeds, but rather to kill the tiny seedling with heat. Properly flamed weeds should have a matt finish on the leaves and pressing your thumb and forefinger together on a leaf should leave a fingerprint. Weeds must have fewer than two true leaves for best burning efficiency. Best results are obtained in windless conditions, in the early morning or evening. Do not use flaming around young trees because it may damage the thin, green bark.

Animals

Weeder geese have been used for years for weed management for various crops and sometimes in orchards. All types of geese will graze weeds. Geese prefer grass species and will eat other crops only after all grasses are gone. They appear to have a particular preference for Johnson grass and Bermuda grass, two weeds that can be particularly troublesome in some orchards. They will even dig out most of the stolons and rhizomes. Caution must be exercised not to place them close to grass crops, such as corn, sorghum, or small grains, which they would probably consider as a delicacy. Sheep and miniature sheep will eat almost all weeds down to ground level, which reduces weed competition, but does not eliminate it. Goats are browsers, so they must be carefully managed to protect the trees. The animals need water, protection from the heat during hot days, and protection from various predators. Movable fencing works well to keep them where they should be.

Management of Special Weed Species

Bermuda Grass (*Cynodon dactylon*)

This is a vigorous, warm season perennial. It expands rapidly with an extensive system of rhizomes and stolons, which can form new plants after being cut by cultivation. Seeds also help Bermuda grass expand to new locations.

Johnsongrass (*Sorghum halapense*)

This is also a warm season perennial that spreads from seeds or from rhizomes. It can overtop small trees and is highly competitive for light, moisture, and nutrients.

Yellow Nut sedge (*Cyperus esculentus*)

This is a perennial weed that reproduces from underground tubers. Viable seeds are rarely produced. A single plant can produce hundreds of new tubers in one year, with tubers able to survive for 2 to 5 years in the soil. Each tuber contains several buds, with each bud capable of producing a plant. Generally only a single bud grows from a tuber, but if a tuber is damaged by cultivation, a new bud is activated. Repeated cultivation at 3-week intervals destroys successive flushes of nut sedge and eventually kills the tuber.

Field Bindweed (*Convolvulus arvensis*)

This is a vigorous perennial weed. It grows from seeds as well as from rhizomes and extensive, deep roots. It has been known to survive over 30 years. Because of the longevity of the seed in the soil it is imperative to destroy the plants before they can produce seeds. The plants may spread from stem or root sections that are cut during cultivations, but cultivation controls seedlings. Repeated cultivation at 2- to 3-week intervals depletes the carbohydrates in the root system and eventually kills the weed. Longer periods between cultivation would allow the energy reserves to replenish.

15

Water Management

Olive trees are considered one of the drought tolerant species which can survive under water stress during tree growth and development periods and a little water support tree longevity. They also show a high capacity to recover from prolonged drought periods. Trees can completely re-hydrate within 3 days of irrigation after a water deficit that reached a leaf water potential (LWP) of -4.0 MPa. Even during a severe drought that lowered the LWP down to -8.0 MPa, the trees rehydrated in less than a week following the onset of rains. However, larger growth of olive leaves, branches, fruit, and trunk, is sensitive to water deficits. Physiologically, water stress regulates stomatal opening and photosynthesis. The stomata close partially during the day in response to increases in vapour pressure deficit, even if, the trees are well watered, with corresponding decreases in CO_2 assimilation. Leaf photosynthetic rate is relatively high, of the order of 12-20 µmol CO_2 $m^{-2}s^{-1}$, but is reduce under water stress because of stomatal closure. When water deficit is severe (LWP below -4 to -5 MPa), photosynthetic rate do not exceed one-third of normal value, reaching a maximum early in the morning and then declining as the day advances.

Olive passes through different phenological stages: leaf development, shoot development, inflorescence development, flower bud induction, flowering, fruit development, pit hardening, ripening and oil accumulation (Figure 6). The fruit and oil yield is the result of three main developmental stages that occur between flowering and harvest: fruit set, fruit growth and oil accumulation in the fruit pulp (mesocarp). Vegetative growth is critical in terms of olive fruit production, because flowering and fruit set originate in the auxiliary buds of past year's growth. The reproductive cycle from flower bud induction to fruit ripening takes 15-18 months depending on cultivar and growing conditions, as it starts in the summer and ends in the autumn of the following year. Flower usually borne in inflorescence at the axial of leaves of one-year old wood; whereas, the terminal bud of the shoot is almost always vegetative. Shoot growth start with bud break in spring and resumes when temperature is above 12 °C, as long as it is not inhibited by temperature above 35 °C, soil water deficit or other environmental stresses. A second flush of shoot growth is common after the summer. Olive trees are sensitive to water logging and temperatures below -10 °C.

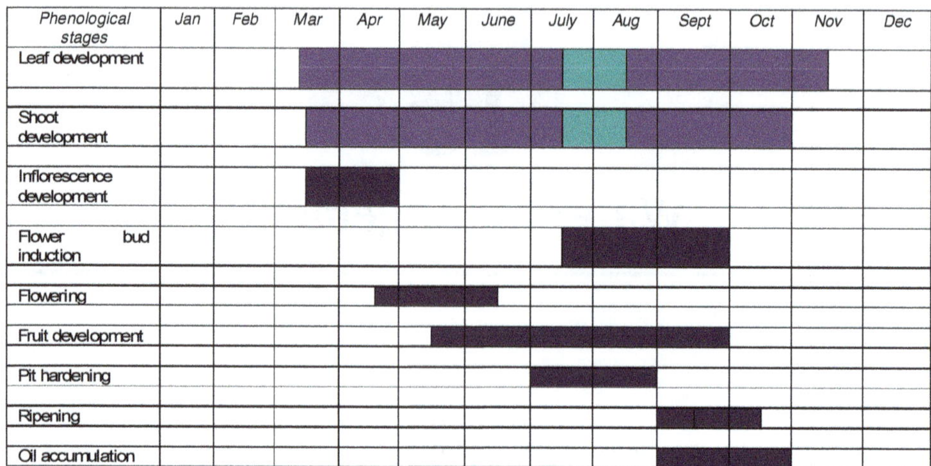

Figure 6: Phenological Stages in Olive Tree and Fruit Growth and Development.

In table olive production, maximum fruit size and fruit yield are essential, while in olive oil production, oil yield and quality are needed for the viability of olive growing. Therefore, the application of adequate water at crucial phenological stages decides orchard productivity since olive is extremely responsive to irrigation in terms of maximizing shoot growth, fruit size, and fruit yield, and oil yield.

Water Use Pattern

The availability of water in the olive production play critical functions through regulation of physiological and biochemical activities related to the growth and development. Therefore, good knowledge of the olive tree in general and specifically the physiological processes are useful for a judicious irrigation management in olive orchards. However, before any sort of regulated deficit irrigation, strategy can be made about the timing and amount of water needed to the trees. Water is lost from olive orchard by evaporation as well as transpiration. This is affected by numerous factors, including humidity, temperature, wind, solar radiation (day length), and the per cent canopy cover (the percentage of ground shaded by trees). Several approaches have been suggested aiming the water productivity as under-

The FAO Coefficient Approach

Determining the crop evapotranspiration (ETc) for non-limited available water in the soil is crucial for calculating irrigation amount (IA). Likely, that of the crop coefficient recommended by the FAO is the most widely used approach for determining *etc*. This is calculated from the potential evapotranspiration (ETo) in the area, a coefficient *Kc* called the crop coefficient and a coefficient *Kr* related to the percentage of ground covered by the crop: ETc = *Kc Kr* ETo. The most widely accepted methods for determining ETo are based on the use either of the evaporation tank or automatic weather stations for recording the variables required for calculating ETo from a combination equation appropriate for the area. The main limitation of the FAO method comes from the empirical character of the *Kc* and *Kr* coefficients: both of them depend on the orchard conditions.

Recent Coefficient Approach

Testi *et al.* (2006) and Orgaz *et al.* (2006) have proposed a model of olive water requirements, which estimates transpiration (*E*p) and soil evaporation (*E*s) separately, and a new crop coefficient *K*c = ETc/ETo, calculated as the sum of three main components: tree transpiration (*K*p), evaporation from the soil (*K*s1) and evaporation from the areas wetted by the emitters (*K*s2). A fourth component can be added, accounting for evaporation of the water intercepted by the canopy (*K*pd).

Water Balance

Determining the components of the water balance equation in the olive orchard is a suitable approach for estimating the fractions of the supplied water used by the crop, stored in the soil or lost by drainage and runoff. Palomo *et al.* (2002) used this approach for two years and three water treatments, in a mature 'Manzanilla' olive orchard. The information obtained was useful not only to quantify the crop water needs, but also to evaluate water losses by drainage, which is important to evaluate the environmental impact of fertigation and to evaluate the irrigation management. This is, however, a labour and time consuming approach, mostly used with research purposes rather than for optimizing crop water use in commercial orchards.

Average Irrigation Requirements

Average ETc and IA values for mature olive orchards in areas with ETo is ranges from 1000 to 1400 mm year^{-1}. For average weather conditions in the Mediterranean basin (ETo 1200 mm year^{-1}, rainfall 500 mm year^{-1}) and mature olive orchards with 100-300 trees ha^{-1} and localised irrigation, maximum potential ETc could be 6000-7000 m^3 ha^{-1} year^{-1}, from which 3000-4000 m^3 ha^{-1} must be applied by irrigation. These are average figures, being necessary to adjust IA for each orchard depending on plant density, canopy volume and characteristics of the irrigation system, among other factors.

Water Losses by Soil Evaporation

Annual *E*s are quite high in olive orchards. Testi *et al.* (2006) determined *E*s to be 40 per cent of the annual ETc in a typically Mediterranean traditional olive orchard (100 trees ha^{-1} at 10 m X 10 m spacing), and 35 per cent in an intensive orchard (300 trees ha^{-1} and individual tree canopy volume of 50 m^3). For the two mentioned cases, annual *E*s from the ground spots wetted by the emitters amounted to 11 and 10 per cent, respectively.

Modelling

Continuous improvements on the understanding of the soil-plant-atmosphere relationships in olive orchards are leading to models that can be used as tools for optimizing irrigation. This is the case of the photosynthesis model for olive leaves published by Diaz-Espejo *et al.* (2006); combined with a model of radiation transfer through the canopy, it could predict the response of the whole-tree carbon assimilation to water stress. Now, the use of these highly mechanistic models to optimise production and the use of irrigation water in commercial orchards is limited by the required number of parameters and variables –some of them difficult to measure. A more practical approach is to use models in which some inputs are estimated from measurements of related variables made in the orchard in which are going to be applied.

Deficit Irrigation

Water for irrigation is scarce in most olive orchards. Therefore, the orchardist is very often bound to apply deficit irrigation (DI) approach. The aim is to applied IA below the crop water needs but in a rational way, to keep the crop performance as close as possible to its maximum potential. The old practice of applying just one or very few irrigation events on the dry season has been called supplementary or complementary irrigation. Sustained DI is when a reduced percentage of ETc is applied all throughout the irrigation season. Low frequency DI is when the soil is left to dry until the readily available water is consumed; then the soil is irrigated to field capacity and left to dry again. Among the most widely used DI approaches are the regulated deficit irrigation (RDI) and the *partial rootzone drying* (PRD). Deficit irrigation strategies are not recommended for young orchards, since conditions for the trees reaching maturity as soon as possible must be favoured.

The olive tree is a parsimonious water consumer well adapted to xeric conditions. It's mechanisms for drought tolerance make the species to be particularly suitable for DI. Benefits of DI on oil quality have been already mentioned. The orchardist must keep in mind, however, that any DI strategy may reduce the crop performance in subsequent years. The success of complementary irrigation, sustained DI or low frequency ID depends on the soil water holding capacity, which must be characterized before any of these strategies is applied.

Regulated Deficit Irrigation (RDI)

This is one of the most widely adopted DI strategies, based on supplying some 100 per cent of ETc when the crop is less tolerant to water stress and a reduced percentage (30 per cent is quite common) for the rest of the season. For the whole season, IA (Irrigation amount) reductions of about 50 per cent of ETc are easily achieved. A detailed knowledge on the tree physiology is crucial for the success of RDI. Currently, in fact, many studies on RDI are oriented to better establish the periods for IA reduction.

Partial Root Zone Drying (PRD)

This is a relatively new DI approach. The aim is to irrigate with similar IAs than in RDI but achieving a greater crop performance. This is achieved by irrigating half of the root zone while the other half is kept under drying soil, alternating irrigation from one half to the other every 2-3 weeks. In theory, this triggers a root-to-shoot signalling mechanism that induces stomata closure and improves water use efficiency. On the other hand, the irrigation system for PRD is more expensive than for a traditional localised irrigation system, since two laterals per tree row are required, and the management is more complicated. Although IA amounted to 50 per cent only of the control treatment irrigated on both sides with 100 per cent of ETc, relative water content and photosynthetic capacity were similar in both treatments, and yield was reduced in 15-20 per cent only, with the same yield quality. Unfortunately they did not have a companion RDI treatment, so doubts remain on whether similar benefits could have been obtained with RDI. We have recently published an experiment in which 50 per cent of ETc was supplied by irrigation to mature 'Manzanilla' trees following both and RDI and a PRD approach.

Irrigation Scheduling

There is a need for a more accurate scheduling of water application in olive orchards: *first*, because of the increasing demand from competing water consumers; *second*, because of the increasing demand for quality, which, in the case of orchards for oil production, may require an accurate scheduling with IAs lower than those required for full irrigation. The techniques for estimating ETc mentioned above are, in some cases, too coarse. Consequently, increasing efforts are being put into the development of new techniques for a more precise irrigation in olive orchards. Basically, water supplies in the orchard can be scheduled from the soil water status, from the atmospheric demand or from plant-based measurements. Among the last ones, leaf or stem water potential, trunk diameter variations and sap flow records are being evaluated by several research groups. Main advantage of plant-based indicators is that the tree is used as a biosensor which responses to the soil water status, the plant characteristics and the atmospheric demand. Irrigation scheduling based on the atmospheric demand has been considered in the first part of this work, so it is not considered here. Finally, infrared thermography is becoming a promising tool to account for orchard variability.

Scheduling Irrigation from Soil Water Measurements

There is a variety of instruments for measuring and soil metric potential (h, MPa). Measurements of any of these variables, together with an adequate hydrodynamic characterization of the soil orchard, can be useful to monitor the soil water status for scheduling irrigation. Some of the sensors available in the market although not very precise are relatively inexpensive and can be easily automated. Main limitations of this approach are the high number of sensors that may be required to have representative measurements, and the fact that neither physiological features of the plant nor the atmospheric demand are taken into account.

Irrigation Scheduling from Plant Water Potential

Measurements of leaf water potential to monitor the response of the tree water status to irrigation have been widely used, with interesting practical results. This technique has several limitations. First, measurements have to be manually made, which is labour consuming and restricts the number of replications. Second, there are uncertainties on the thresholds to be used.

Irrigation Scheduling from Trunk Diameter Variations

Continuous monitoring of trunk diameter changes by linear variable displacement transducers (LVDT sensors) for assessing the tree response to irrigation water deficits has been evaluated for a variety of fruit tree species. The fundamentals of the technique are described by Goldhamer and Fereres (2001). In young olive trees, results are difficult to interpret due to the influence of trunk growth.

Irrigation Scheduling from Sap Flow Measurements

Both olive water consumption and the dynamics of transpiration and water uptake by main roots can be estimated from sap flow measurements. Comparisons between sap flow and trunk diameter readings, as water stress indicators in fruit trees, and between these two variables and more traditional water stress indicators such us leaf or stem water potential and stomatal conductance.

Infrared Thermography

This technique has some disadvantages that must be overcome before becoming widely adopted by growers. Thus, being capable to detect the right time for irrigation, does not indicate the amount of water need it. In addition, is still expensive, and image analysis requires sophisticated software. On the other hand, however, the technique allows to characterize variability within the orchard due to both differences in soil and crop conditions and problems with the irrigation system; it performs well in hot and dry conditions, typical for most olive orchards; and it is suitable for large cropped areas. These advantages makes infrared thermography, combined with measurements at the orchard level, to be considered by many as the most promising approach for irrigating commercial orchards in a rational way.

References

Díaz-Espejo, A., Verhoef, A., Fernández, J. E., Villagarcía, L. 2004. Use of high resolution weighing lysimeters to estimate main driving variables related to soil evaporation in a drip irrigated olive orchard. In: Proc. of the 5th Internacional Symposium on Olive Growing. 27 September – 2 October, Izmir, Turkey, pp. 30.

Goldhamer, D.A., Fereres, E. 2001. Irrigation scheduling protocols using continuously recorded trunk diameter measurements. *Irrigation Science* **20**(3): 115-125.

Orgaz, F., Testi, L., Villalobos, F.J., Fereres, E. 2006. Water requirements of olive orchards-II: determination of crop coefficients for irrigation scheduling. *Irrigation Science* **24**: 77-84.

Palomo, M.J., Moreno, F., Fernández, J.E., Díaz-Espejo, A., Girón, I.F. 2002. Determining water consumption in olive orchards using the water balance approach. *Agricultural Water Management* **55**: 15-35.

Testi, L., Villalobos, F.J., Orgaz, F., Fereres, E. 2006. Water requirements of olive orchards: I simulation of daily evapotranspiration for scenario analysis. *Irrigation Science* **24**: 69-76.

16

Canopy Management

Growing tree skeleton is also known as the tree canopy, consisting of tree trunk, primary, secondary, tertiary branches and the fruiting wood and their respective location on the tree. The canopy architecture is the systematic distribution, orientation of branches; size and shape of the plant or organs such as leaves, stems, branches and flowers. The type of canopy architecture therefore essentially affects crop functions such as light interception (LI) and evapo-transpiration and eventual biomass accumulation and fruit production. Canopy architecture also includes the elements like leaf morphology or shape; tree growth habit (*i.e.*, individual plant shape) and number of nodes on main stem as a proxy for number of primary branches. Therefore, the management of these plant parts are required to facilitate the fruiting organs to their potential. Modification in the canopy architecture affects fruit quality and yield by affecting light interception. Therefore, the canopy geometry should be managed in such a way that it can intercept maximum light. For high density, a shallow canopy of 1.5 to 2.0 m in depth is needed to achieve the maximum efficiency for trapping sun energy through foliage and channelizing metabolites for quality fruit production with good return.

Light is an important aspect of canopy studies, because of its role in photosynthesis and its need in development of morphology of leaves and shoots, role in flower initiation, fruit set, yield and quality. The objective of canopy management is to optimize the plant model in such a form so that it may intercept maximum light by tree training, pruning, branch and tree orientation, use of growth regulating chemicals, planting space and design. Interception of light by total orchard canopy is correlated to total biomass production of the crop. Light interception and distribution within a single tree canopy must be optimized harnessing of tree genetic constitution with regard to yield and quality. Shikhamany (2001) described basic concept of canopy management for efficient utilization of light.

1. Maximum utilization of light.
2. Avoidance of the build-up of micro-climatic congenial for the disease and pest.
3. Convenience in carrying out the cultural operations.
4. Maximizing the productivity and quality.
5. Economy in obtaining the required canopy architecture.

Olive trees store most of their energy in their leaves, and unlike deciduous trees, do not show much response to pruning. The pruning that removes large amounts of foliage can stunt olive trees just like summer pruning which removes leaves on a deciduous tree. The removal of too many leaves and shoots can drastically reduce the availability of assimilates in young plants with small photosynthetic area. Moreover, severe pruning of young trees decreases the shoot to root ratio and induces a strong vegetative response that delays the onset of production. Pruning olive trees is quite different from pruning other fruit trees of the temperate zone, because of their biological peculiarities. Errors in pruning may result in yield losses or higher cultivation costs. Pruning also determines the training system which, in turn, is one of the major factors for successful tree performance and orchard profitability.

Consideration in Canopy Management

1. The less pruning during training, the early the onset of production and the higher the yield.
2. The more regular the skeleton structure to be achieved at maturity, the more frequent and intense the pruning should be the first few years after planting.
3. Eliminate only shoots which disturb the definitive shape of the plant; completely eliminate water sprouts and suckers if appear.
4. The lateral shoots competing vigorously with the main stem are to be eliminated to promote the growth of the central leader or of primary branches. Shoots inserted at the same height compete strongly with one another and with the main axis and will seldom develop into strong scaffolds. These shoots must be thinned to avoid their progressive weakening.
5. Pruning should be gradual during the training phase.

There are many ways to prune olive trees, but a few general training forms are widely adopted. Olive trees normally grow in a basal form, which means that lower laterals will grow just as vigorously as the leader. Consequently, a bush is formed rather than a "Christmas tree" shape. The easiest way to manage this tree habit is to allow for the development of 3 or 4 scaffold branches and to keep the center open to maintain good light exposure in the lower portion of the tree. Unless the desire is for a low bush-shaped tree with multiple trunks, the single trunk tree is much preferred. Single trunk trees provide a location for trunk shakers to grab onto if trunk shaking. Multiple trunk trees tend to grow wide into the row and the outermost shoots grow as vigorously as the leader, making it difficult to form the desired tree shape. It is also easier to control weed competition around single trunk trees. Remove all side branches up to 3 feet from the ground. The mini central leader is used in super-high-density systems to maintain a central trunk from which laterals are continuously renewed every three years, but it is a very intensive system and not the natural shape in which olive trees grow.

Operational Guideline for Establishing a Desired Canopy

At Planting

Immediate after the planting all shoots below a 30-inch (75cm) height should be removed and the leader should be pinched at 36 to 40 inches (90 to 100cm) to stimulate lateral growth. If lateral shoots above 30 inches have begun growth on the tree in the nursery, no pinching of the terminal is required.

First Summer

Early in the growing season, select three to five lateral-growing, upright shoots to be the scaffold branches. Space the scaffolds around the trunk, leaving sufficient distance between them to allow eventual shaker access. One to two-inch (2.5 to 5cm) vertical separation between scaffolds is necessary to provide strength on attachment to the trunk. Remove any other vigorous shoots on the trunk while they are quite small to direct growth into the selected scaffolds. Avoid larger cuts late in the growing season, since they would delay production.

Second and Third Growing Seasons

The only pruning suggested for the second and third growing seasons serves to continue directing tree growth into the framework branches. Usually, all that is required is the early removal of suckers, water sprouts, and excessively low-hanging shoots. Once bearing begins, ideally in the third growing season, a secondary scaffold can be selected and developed.

Generally open vase and modified open vase pruning styles are used to maximize fruit production and fruit quality for fruit trees in the genus *Prunus* (peach, nectarine plum apricot and their interspecific hybrids; also almond) as nectarine, plum, apricot and their interspecific hybrids; also almond) as well as for olive trees.

Pruning Olive Tree

At the time of planting, a support system of about 10 feet should be erected. In a feathered tree, develop a clean stem to a height of 45 cm from the ground and select one branch each to the left and right which can be tied to lowermost wire. Remove weak laterals. In the second growing season, develop another set of scaffolds on the next higher wire and repeat the process. Eliminate all other undesirable scaffold branches. In third growing season, another pair of lateral scaffold branches is developed to the next higher wire. The secondary scaffold branches are allowed to develop on main scaffold branches but if they become over-vigorous, they should be headed back. All upright growing shoots which either compete with the leader branch or arise from main scaffold limbs are eliminated soon after their emergence.

When tree attains a height of 4 m, it is headed back to a weak lateral. The intensity of pruning in later years is kept the minimum until tree attains bearing age.

Young non-bearing trees can be pruned any time of the year if weather is clear and free from intermittent rains. However, pruning of trees should be avoided when prolonged drought period coupled with acute water stress prevails. In bearing trees, pruning should always be carried out immediately after harvesting to encourage new growth for initiating floral buds for the next season.

The intensity of pruning depends upon vigour of cultivar, age of tree, availability of irrigation, and bearing behaviour of the tree. During transitional phase, only light pruning is adopted but as the tree becomes older, the intensity of pruning is to be increased accordingly. In old trees, rejuvenating type of pruning is required which involves heading back of main scaffold limbs to encourage new vigorous shoots which begin fruiting after 2 years.

In young plants, only corrective pruning should be done as and when necessary. Pruning during initial 2–3 years is not advisable. However, occasional removal of water sprouts, dead or mechanically injured branches should be carried out regularly. A clean trunk devoid of any feathers should be developed up to a height of 40cm above the ground. The higher intensity of pruning during initial years stimulates strong vegetative growth. The pruning wounds should be immediately disinfected with Bordeaux paint. This minimizes the chances of various pathogens entering into the plants and also hastens the process of healing.

Training Systems for Higher Productivity and Better Fruit Quality

Bush Form

It is the best form (Figure 7) for manual harvesting. No pruning is required soon after the time of plantation for this system, because the natural behaviour of olive trees. This is the way to obtain early bearing and high yield from the tree in a few (three to five) years. This form can remain untouched for some years and occasionally it may be pruned only the inner branches including the water sprouts which may allow the tree to grow upright. Some problems for this tree form may be seen after some years (from 8 to 15) because in the lower part of the canopy branches start losing their vigour due to overlapping and criss-crossing of twigs; the tree is more affected by pest and diseases. Nevertheless the top portion of the canopy tries to re-place the lower weak portion; the yield decreases thereby increasing production costs. To avoid these problems it is necessary to a) select cultivars having less tree vigour, b) go for low density plantations, c) find suitable environment, and d) head-back the main branches with severe cuts,

Open Centre System (vase-with a cylindrical or truncated-conical top)

This form (Figure 8) is better for manual harvesting with combs working with air pressure. In this form, we delay the beginning of fruits production. The canopy is having a trunk of 80 to 120 cm and three to four main branches having conical shape. The object of this form is to give every part of the tree a large and equal amount of light and air. To obtain this form, we select three to four well placed

Figure 7: Tree Trained to a Bush Form.

new shoots (which will form the main branches) and tie them to a tutor (bamboo sticks) having 45° inclination. We can avoid to tie the selected shoots if we do; a) At plantation time the tree maiden is tied to a support or stake in a position that we find suitable for the main branches; b) During the growing period the maiden will bend alone due to its weight (if not is better to pinch-out the tip) and many new and vigorous shoots are formed at the bending point; c) All coming shoots must be well checked and need to prune the internal shoots (shoot thinning), water shoots and need instead to keep the shoots bending towards the external part of the tree;

Figure 8: Open Centre of Trees in Young Olive Orchard.

d) In between the selected shoots, we choose three to four well placed branches for the formation of the main branches; e) In two to three years the trunk will be formed; then it is necessary to remove the shoots arising along the stem to avoid competition for nutrients, space, *etc.*; f) Annual pruning operation (the tree structure is formed with a maximum height of 5 m) is done to allow the tree to increase in volume but not in height, expanding its canopy along with the side laterals of each main scaffold branch.

Low Headed Open Centre System

This form is almost like the previous one (Vase) with the exception that the trunk is very short. The tree at the time of plantation is headed- back to 30- 50 cm height and those shoots originated near the cut will be selected for the formation of the tree structure. A special care is required for developing shoots, as in the open centre form. This is a half-way form between the bush and the open centre; so it carries some problems of the bush (partially) and the requirements/benefits of the open centre (Figure 9).

Figure 9:Low Headed Open Centre Trees in Olive Orchard

Monocono System (Spindle Type of Free)

This form (Figure 10) is related to modern olive orchards where mechanical harvest method is required and high density plantation is followed. The olive is trained to a single axil, with lateral branches decreasing in size from the bottom to the periphery, with a portion of trunk (60 to 80 cm and close to the soil) free from shoots or branchlet formation to permit the operation of the shaker. Since the fruits are placed relatively near the axil, the detachment of the stalk from the twig is easier at the time of shaking the tree. This form however may have several problems, such as:

Figure 10: Monocono System (Central leader system).

1. Cultivars having spreading or drooping habit are difficult to train.
2. It requires constant and qualified pruning operations.
3. It requires long stakes or supports and wires.
4. Trees become too high and is difficult to control.
5. There is competition between the canopies because they try to touch eachothers.

Espalier Training System

Espalier training offers several benefits for the grower. Trees trained to grow flat against a stone wall save space in the garden. Training a tree lower to the ground allows for better access, making care of it easier. An espaliered olive tree is easier to prune and pick. The trees act as a focal point in the garden, as an espaliered olive tree can take on different forms- espalier designs include 'candelabra', 'double U-shape', 'Belgian fence', 'Belgian arch', 'Belgian doublet' and 'fan shape'. Depending on the wall construction, various tools are needed to help train the olive trees and secure branches. Some common tools for espalier include 12- to 15-gauge wire; double-pointed staples for wood wall or fencing; and wall mounts for masonry, including U-bolts and eye screws. Biodegradable binding tube often is used to train trees into the desired pattern as well. Your local nursery, garden centre or hardware store likely carries various types of anchoring devices. No Matches Found. Please try your search again. Espalier olive training is an on-going process that requires regular pruning and training to obtain and maintain a specific shape. Branches that grow outward, away from the intended pattern, need pruning. Branches used to create the intended design require ties to keep them in place. Prune olive trees in spring or summer and before flowering.

References

Shikhamany, S.D. 2001. In: Grape production in the Asia Pacific Region by Food and Agriculture Organization of the United Nations Regional Office for Asia and the Pacific Bangkok, Thailand.

Wells and Norman, 1991.Instrument for indirect measurement of canopy architecture. *Agron. J.* **83**: 818–825.

17

Use of Plant Growth Promoting Substance

Plant growth regulators or phytohormones are organic substances produced naturally in higher plants, controlling growth or other physiological functions at a site remote from its place of production and active in minute amounts. Thimmann (1963) proposed the term *Phyto hormone,* as these hormones are synthesized in plants. *Plant growth regulators* include auxins, gibberellins, cytokinins, ethylene, growth retardants and growth inhibitors.

Plant Growth Regulators

- ☆ Defined as organic compounds other than nutrients, that affects the physiological processes of growth and development in plants when applied in low concentrations.
- ☆ Defined as either natural or synthetic compounds that are applied directly to a target plant to alter its life processes or its structure to improve quality, increase yields, or facilitate harvesting.

Types of Plant Hormones

There are five general classes of hormones: auxins, cytokinins, gibberellins, ethylene, and abscisic acid.

Auxins

Auxin as indole-3-acetic acid (IAA), was the first plant hormone identified. It is synthesized primarily in the shoot tips (leaf primordia and young leaves), in embryos, and in parts of developing flowers and seeds. It move from cell to cell through the parenchyma surrounding the vascular tissues requires the expenditure of ATP energy. IAA moves in one direction only—that is, the movement is polar and, in this case, downward. Such downward movement in *shoots* is said to be basipetal movement, and in *roots* it is acropetal.

Auxins alone or in combination with other hormones are responsible for many aspects of plant growth. IAA in particular:

- ☆ It helps in activation of the differentiation of vascular tissue in the shoot apex and in calluses; initiates division of the vascular cambium in the spring; promotes growth of vascular tissue in healing of wounds.
- ☆ Promotes initiation and growth of adventitious roots in cuttings.
- ☆ It increases the plasticity of the cell wall by activating cellular elongation.
- ☆ It helps in maintaining apical dominance indirectly by stimulating the production of ethylene, which directly inhibits lateral bud growth.
- ☆ Activates a gene required for making a protein necessary for growth and other genes for the synthesis of wall materials made and secreted by dictyosomes.
- ☆ Promotes the growth of many fruits (from auxin produced by the developing seeds).
- ☆ Suppresses the abscission (separation from the plant) of fruits and leaves (lowered production of auxin in the leaf is correlated with formation of the abscission layer).
- ☆ Inhibits most flowering (but promotes flowering of pineapples).
- ☆ Controls aging and senescence, dormancy of seeds.
- ☆ Synthetic auxins are extensively used as herbicides, the most widely known being 2,4-D and the notorious 2,4,5-T, which were used in a 1:1 combination as Agent Orange during the Vietnam War and sprayed over the Vietnam forests as a defoliant.

Cytokinins

Named because of their discovered role in cell division (cytokinesis), the cytokinins have a molecular structure similar to adenine. Naturally occurring **zeatin,** isolated first from corn (*Zea mays*), is the most active of the cytokinins. Cytokinins are found in sites of active cell division in plants for example, in root tips, seeds, fruits, and leaves. They are transported in the xylem and work in the presence of auxin to promote cell division. Differing cytokinin:auxin ratios change the nature of organogenesis. If kinetin is high and auxin low, shoots are formed; if kinetin is low and auxin high, roots are formed. Lateral bud development, which is retarded by auxin, is promoted by cytokinins. Cytokinins also delay the senescence of leaves and promote the expansion of cotyledons.

Gibberellins

The gibberellins are widespread throughout the plant kingdom, and more than 75 have been isolated, to date. Rather than giving each a specific name, the compounds are numbered—for example, GA_1, GA_2, and so on. Gibberellic acid three (GA_3) is the most widespread and most thoroughly studied. The gibberellins are especially abundant in seeds and young shoots where they control stem elongation by stimulating both cell division and elongation (auxin stimulates only cell elongation). The gibberellins are carried by the xylem and phloem. Numerous effects have been cataloged that involve about 15 or fewer of the gibberellic acids. The greater number with no known effects apparently is precursors to the active ones. Experimentation with GA_3 sprayed on genetically dwarf plants stimulates elongation of the dwarf plants to normal heights. Normal height plants sprayed with GA_3 become giants.

Ethylene

Ethylene is a simple gaseous hydrocarbon produced from an amino acid and appears in most plant tissues in large amounts when they are stressed. It diffuses from its site of origin into the air and affects surrounding plants as well. Large amounts ordinarily are produced by roots, senescing flowers, ripening fruits, and the apical meristem of shoots. Auxin increases ethylene production, as does ethylene itself—small amounts of ethylene initiate copious production of still more. Ethylene stimulates the ripening of fruit and initiates abscission of fruits and leaves. In monoecious plants (those with separate male and female flowers borne on the same plant), gibberellins and ethylene concentrations determine the sex of the flowers: Flower buds exposed to high concentrations of ethylene produce carpellate flowers, while gibberellins induce staminate ones.

Abscisic Acid

Addicott *et al.* (1965) isolated a substance strongly antagonistic to growth from young cotton fruits and named Abscissin II. Later on this name was changed to Abscisic acid. This substance also induces dormancy of buds therefore it also named as Dormin. Abscisic acid is a naturally occurring growth inhibitor.

Growth Retardants

There is number of synthetic compounds which prevent the gibberellins from exhibiting their usual responses in plants such as cell enlargement or stem elongation. So they are called as anti gibberellins or growth retardants. They are:

1. Cycocel (2- chloroethyl trimethyl ammonium chloride (CCC)
2. Phosphon D – (2, 4 – dichlorobenzyl – tributyl phosphonium chloride)
3. AMO – 1618
4. Morphactins
5. Maleic hydrazide

Uses in Olive Production

1. Root Initiation

Significant efforts already have been made to produce high value, true-to-type, and disease free nursery plants, but olive stem cuttings are hard to root under ordinary conditions. Plant growth regulators can play a vital role to enhance rooting of olive cuttings. Indole butyric acid is an auxin which is commonly used to promote the formation of roots in plants and to generate new roots in the cloning of plants through cuttings (Weisman *et al.* 1995). Indole-3-butyric acid (IBA) treatment improves rooting of easy-to-root and moderately hard to root olive cultivars, but has less effect to stimulate root formation in hard to root cultivars (Ahmed *et al.* 2010). Various studies have revealed that treatments with growth regulators have to a great extent influenced the rooting ability of olive cuttings. The propagation of olive cuttings under mist has become a usual practice throughout the world. Some olive cultivars are easily rooted while others are not. The semi hardwood cuttings treated with 3,000 mg.L^{-1} IBA showed increased percentage of rooting and average number of roots. Similarly the application of 4000 ppm IBA was found the best and to obtain the most number of rooted cuttings, application of 3000 ppm IBA was the best (Kurd *et al.*, 2010).

As an alternative to the traditional use of rooting media, shoots may be dipped in to a liquid rooting solution and inserted directly in to an auxin free mediun. Obviously, the auxin concentration is markedly higher in the dpping method, in comparison with the common system of using a rooting medium. Rugini and Fedeli (1990), for instance, reported high rooting percentage by dipping entire microcuttings, or only their basal parts, in an IBA solution (100-200 mg/L) for 10-20 sec. Because of its laboriousness, it is advisable to try the dipping method when gelled rooting media do not produce acceptable and consistent results.

2. Micro-propagation

The inability to induce adventitious roots is often a limiting factor in conventional cuttings and tissue culture. Several successful attempts has been made in the rooting of micropropataed olive shoots over the last decades, so that even cultivars 'recalcitrants ' to cutting propagation can now be satisfactory rooted *in vitro*. Auxin, known to be involved in cell enlargement, was long thought to be the controlling factor in the rooting process. Two types of evidence support this reasoning: (a) the increased content of endogenous auxin in the base of cutting during rooting induction and (b) the rooting response of many plants to exogenous auxin. Success in micropropagation is thus mostly dependant on the production of good quality adventitious roots, whose formation has three distinct phases' viz., induction, initiation and expression. Medium formulation and auxin treatments are both important factors to promote olive shoot rooting. As regards root promoters, both IBA and 1-naphthaleneacetic acid (NAA) have been indicated as most effective for the stimulation of rooting. IBA is supplemented at concentration of 2-3 mg/L. As NAA produces a stronger stimulation than IBA, it is most often applied at 1 mg/L concentration. In comparison with cytokinin, IBA and NAA are more stable when heated at 121°C. Therefore, they can be added to the rooting medium before autoclaving with no need to increase the concentration.

The indolebutyric acid (IBA) and Naphthaleneacetic acid (NAA) hormones are known for higher percentage of rooted shoots, root number, root length and quality of roots, IBA at 1.5 mg l^{-1} concentration and proved to be the best one for rooting of 'Moraiolo' cultivar of olive. The roots produced on IBA were longer with better quality shoots, whereas, NAA produced poor response with necrotic leaves and leaf abscission.

3. Fruit Set

Some of the main cause of inconsistent productivity in olive is self incompatibility, poor and slow pollen germination and an abnormal embryo sac determining ovule degeneration. Production control by means of phytoregulators and biostimulents, which are widely used on deciduous fruit trees, has proven to be more complex and harder to interpret in olive an evergreen species with a greater number of metabolic sink during the flowering stage. In one of the experiment in olive cv. Frontoio was treated before and during bloom with several bio-stimulants and found that treatment leads to increase in fruit set and plant productivity with positive effect on several characteristics of the stone fruit. Similarly, GA_3 application @ 30 ppm caused more pronounced effect on number of inflorescence, flower count, perfect flower percentage, fruit drop percentage, fruit set percentage, fruit harvest percentage, fruit size, fruit weight, pulp weight and oil contents (Khalil *et al.*, 2012)

4. Reduction in Alternate Bearing

Application of Gibberellic acid (GA_3) before an expected "on" year decreases the percentage of opened flower buds per shoot, number of flowers per inflorescence in the subsequent year. GA is responsible for cell elongation, rather than cell division (Francis and Sorrell, 2001). GA_3 succeed in inhibition of olive tree blooming in the following season (El-Sharkawy, 1999). GA_3 has the potential control on growth and flowering process. In addition, increase petiole length, leaf area and delay petal abscission and color fading (senescence) by the hydrolysis of starch and sucrose into fructose and glucose (Khan and Chaudhry, 2006); GA_3 treatments increase the weight, length, width of olive fruits, stone weight and flesh weight than untreated ones, (Rotundo and Gioffre, 1984); improves fruit weight and flesh: seed ratio, (Ramezani and Shekafandeh, 2009). Naphthaleneacetic acid has important role in fruit formation, abscission cell elongation, apical dominance, photoperiod and geotropism (Haidry *et al.*, 1997). NAA application at (100, 150) mg L^{-1} 15 days after full bloom has been used to chemically thin olives in various countries (Lavee, 2006).

The GA_3 spray before an expected "on" year decreases the percentage of opened flower buds per shoot, number of flowers per inflorescence in the subsequent year. GA_3 succeed in inhibition of olive tree blooming in the following season (El-Sharkawy, 1999). However, Fernandez-Escobar *et al.* (1993) injected non bearing olive trees with GA_3 at 1000 ppm between May and November and mentioned that flowering was inhibited when GA_3 was injected into olive trees and this inhibitory effect varied with timing. Also, Boulouha *et al.* (1993) found that, spray olive trees with GA_3 increase the annual vegetative growth in long and short fruit bearing branches. Also, the treatment of olive trees with GA_3 in July increase the vegetative growth in both the treatment year and the following year (Proietti and Tombesi, 1996). In spite of GA_3 application decreased the following yield and reduced the fruiting (Hassan, 1987 and Lavee and Haskal, 1994) but it increase fruit yield of the following year and appeared to stabilize fruiting.

5. Role in Fruit Abscission/Thinning

Various ethylene-releasing compounds (ERC's) are used in order to increase fruit removal and improve mechanical harvesting. In a study on two cultivars ("Frantoio" and "Leccino") The Ethrel treatment was applied at 2000 ppm and various abscission zones were observed that are differently activated during fruit development.

6. Yield and Quality

The growth regulators have clear cut significant positive effect on fruit drop, yield and quality of olive. In an individual study when thirteen years old olive trees were foliar sprayed with both GA_3 and NAA individually and additively at 50 and 75 ppm. The maximum fruit drop per cent was recorded by using NAA at 75 ppm. Spraying trees with 75 ppm of GA_3 and NAA at either 50 or 75 ppm decreased fruit drop per cent in comparison to those sprayed with NAA only. Spraying trees with GA_3 at 75 ppm caused maximum fruit yield/tree in comparison to those of other treatments including control. Maximum fruit weight, volume, length, diameter as well as fruit shape index were obtained when trees were treated with GA_3 at 75 ppm. GA_3 and NAA either individually or additively increased TSS per cent as well as TSS/Acid ratio of fruit juice and decreased total acidity than control. Maximum oil content (per cent of dry wt.) was recorded when trees were sprayed with 75 ppm of GA_3 in comparison with other treatments including control. Accordingly advisable to spray olive trees with GA_3 and NAA individually or additively 10 days after fruit set to improve tree yield and fruit quality.

References

Addicot, F.T., Smith, O.E., Lyon, J.L. 1965. Some physiological properties of abscisin II. *Plant Physiol.* **40** : Supple. XXV.

Ahmed, A.K., Amanullah, S. K., Basharat H. S. and Munir, A. K.2010. Effect of indole butyric acid (IBA) on rooting of olive stems cuttings. *Pakistan J. Agric. Res.* **23**: 3-4.

Bartolini, S., Viti, R. and Vitagliano, C. 1993. Effects of different growth regulators on fruit-set in olive. *Acta Hort.* (ISHS) **329**: 246-248.

Boulouha, B., Wallali, L.D., Loussert, R., Lamhamedi, M. and Sikaoui, L. 1993. Effects of growth regulators on growth and fruiting of Olive (*Olea europaea* L.).Actions de certains phytohormones sur la croissance et la fructification de l'olivier (*Olea europaea* L.). *Al Awamia* **70**: 74- 96..

El-Sharkawy, S.M. 1999. Physiological studies on flowering and fruiting of olive tree. Ph.D. Thesis, Fac. Agric., Moshtohor Zagazig Univ., Egypt, pp: 168.

Fernandez-Escobar, R. and Benlloch, M. 1992. The time of floral induction in the olive. *J.Amer. Soc. Hort. Sci.*, **117**(2): 304-307.

Francis, D. and Srrell, D.A. 2001. The interface between the cell cycle and plant growth Regulators: a mini review. *Plant Growth Regul.* **33**: 1-12.

Haidry, G., Jalal-ud-Din, A. and Munir, M. 1997. Effect of NAA on fruit drop yield and quality of mango, *Mangifera indica* cultivars Langra. *Scientific Khyber* **10** (1): 13- 20.

Khan, A.S. and Chaudhry, N.Y. 2006. GA_3 improves flower yield in some cucurbits treated with lead and mercury. *Afr. J. Biotechnol.*, **5**: 149-153.

Kurd, A. A., Amanullah, Saifullah, K., Shah, B. H., Khetran, M. A. 2010. Effect of indole butyric acid (IBA) on rooting of olive stem cuttings. *Pakistan Journal of Agricultural Research* **23**: 193-195.

Lavee, S. and Haskal, N. 1994. Protein content and composition of leaves and shoot bark in relation to alternate bearing of olive trees (*Olea europaea* L.). *Acta Hort* **356**: 143-147.

Lavee, S. 2006. Biennial bearing in olive (*Olea europaea* L.). FAO Network. *Olea* **25**: 5-12.

Proietti, P. and Tombesi, A. 1996. Effects of gibberellic acid, aspargine and glutamine on flower bud induction in olive. (Abstract) *J. Hort. Sci.* **7** (3): 383 - 388.

Ramezani, S. and Shekafandeh, A. 2009. Roles of gibberellic acid and zinc sulphate in increasing size and weight of olive fruit. *Afr. J. Biotechnol.* **8** (24): 6791-6794.

Rotundo, A. and Gioffre, D. 1984. The effect of gibberellic acid (GA_3) on the productivity of two olive cultivars. *Tecnica Agricola*, **34**(3): 187-202.

Thimann, K. 1963. Plant growth substances: past, present and future. 1963. *Ann. Rev. Plant Physiol.* **14** : 1–18.

Weisman, Z. and Lavee, S. 1995. Enhancement of IBA stimulatory effect on rooting of olive cultivar stems cuttings. *Sciencia Horticulturae* **62**: 89-198.

18
Fruit Maturity, Ripening and Harvesting

Maturity at harvest is the most important factor that determines the shelf life and quality of the fruit. Immature fruits are more prone to shrivelling and mechanical damage and are inferior in flavour and quality when ripe. Over mature fruits are likely to be very soft and mealy with insipid flavour soon after harvest. The fruits harvesting too early or too late are more highly susceptible to postharvest physiological disorders. The olive fruits reach their best when allowed to ripen on the trees. Therefore, maturity indices are important for deciding when the olive fruits should be harvested. The fruit maturation in olive depends on variety, temperature, sunlight, and irrigation. A hot fall can cause fruit to ripen quickly. Early winters do not allow fruit to ripen. Therefore, it is desirable to go for forced harvesting at greener fruit and to hedge against frost damage or a big storm. Some of the varieties ripen faster than others, and in some parts of the orchard than in others.

The ripening is the sequence of the processes that occurs in the later stages of the growth and development through the early stages of senescence and that result in characteristic aesthetic, food quality as evident by the changes in the composition, colour, texture and other sensory attributes. The ripening can be observed visually in olive varieties as they gradually change colour. The skin usually turns from deep green to violet and black. The colour and texture of the flesh also change during these stages as do the colour and sensory characteristics of the oil.

The ultimate flavour of any variety can be completely changed by either harvesting the fruit green (unripe) or mature (ripe). The subtleties in between those two extremes still can have a big influence on the style of oil produced. However, fruit maturity has a greater influence on quality than the variety itself. Green oils have the green herbaceous characteristic and riper fruit has more of an olive fruity flavour; while the oil from very ripe fruit is often buttery, less fruity to flat, and does not keep as well. Greener the fruits the more is the bitterness and pungency the ultimate product and the longer its shelf life. Maturity is often a compromise, but is a key factor in determining the style of oil produced. For any given variety there is probably no more than about two to three weeks of ideal harvest period to capture its best qualities. Oil synthesis and accumulation in the olive fruit occurs over about 34 weeks, begins about 10 weeks into the season, increases fairly rapidly up until fruit maturity (colour change and softening) then the rate of accumulation tapers off, but still continues. It seems like there is a much larger increase than there really is late in the fruit-ripening season due to the loss of moisture in the fruit. When the fruit becomes over-ripe, oil synthesis stops completely.

Specific Maturity and Ripening of Olive

Green (Immature) Fruit

Immature olives are green and firm. They produce oil that is bitter and grassy with unripe and vegetative characteristics. These oils are high in polyphenols (anti-oxidants) and other flavour components. As a result, they are quite bitter and pungent and have a long shelf life. The chlorophyll content is high so the oils are often quite green. It is more difficult to extract oil from unripe olives because the oil containing vacuoles within the cells are not easily ruptured. The olive paste needs to be malaxed longer.

Veraison

As the olive fruit matures from green to yellow-green, it starts to soften and then the skin turns red-purple in colour is called *veraison*. The olives contain high polyphenol content at this stage, and are starting to develop some ripe-fruity characteristics. Oils produced from fruit harvested at this stage have some bitterness and some pungency. They have close to a maximum amount of oil per dry weight. The olives are often considered to be at their peak for olive oil production.

Black (Mature) Fruit

In the process of fruit maturation, the skin colour turns from purple to black (although some varieties never turn completely black), and the flesh darkens all the way to the pit. At this stage, the polyphenol and the chlorophyll contents decline and the carotenoid content increase. Therefore, oils produced from late harvest fruit tend to be more golden in colour, less bitter, less pungent, and have a shorter shelf life. They are often described as sweet oils but are high oil yielder.

Effect of Maturity on Oil Properties

The degree of maturity directly associated with the oil quality. During the fruit growing period, the content of polyphenols gradually increases and reaches a maximum level just as the fruit skin begins to change colour (veraison). As the fruit matures with full colouration up to the pit, the content of polyphenols and most of the other flavour components of the fruit decline very rapidly (over about 2-5 weeks). Oil quality, therefore, is very strongly tied to fruit maturity.

Terroir is the influence of the "land" on the quality of the oil. When the oil is good, people often say it is because of the land - the ideal soil, wonderful climate, or because the olives were dry farmed. Of course when it tastes bad, that is blamed on something or somebody else, usually the processor – never the land. There is undoubtedly an influence of climate, elevation, irrigation, soil, water holding capacity, soil mineral content, sunlight intensity, rainfall, *etc.* on ultimate quality of the oil, but these things are very difficult to pin down. In my opinion and the opinion of most scientists, the influence of terroir is minimal compared to variety and fruit maturity.

The best quality oils come from olives matured to at least the red-ripe stage (Table 7). Fully mature, ripe fruit yield sweeter oils with no bitterness and pungency, but during harvest, they are soft and easily damaged. Immature olives that are green or straw colour are sometimes processed because of the unique flavour they impart to the oil. There are many choices to be made depending on the desired type of oil flavour, its shelf life, its healthy components, and its colour. For example, with low polyphenol content varieties such as 'Arbequina' or 'Sevillano', a 1-month delay in harvest can cause as much as a 4-month loss in shelf life. Later harvest usually yields a better percentage of oil per ton of fruit, so processors and growers are often interested in harvesting as late as possible to augment oil quantity. The olive tree manufactures and stores oil in the fruit throughout the season, but the rate of oil storage flattens and stops just before maturity due to the low light intensity and cool temperatures, providing no real gain in oil content. Olives naturally loose moisture in the maturation process, so the perceived rise in oil content late in the season is actually a loss of fruit moisture.

Table 7: The Effect of Olive Fruit Maturity on Various Characteristics of the Oil

Oil Characteristics	*Green*	*Veraison*	*Black*
Organoleptic	Bitter and grassy with unripe and vegetative characteristics	Some ripe-fruitiness. Some bitterness and pungency	Sweet oils
Yield	Low	Close to maximum per dry weight	High
Anti-Oxidants	Highest	High	Lower
Shelf-Life	Highest	High	Lower
Color	More green	Variable	More golden
Ease of Milling	Longer malaxation needed, can be difficult	Normal	Overripe, overwatered fruit can create problems

Olive Maturity Index

In olive, all fruit in a orchard, or even on the same tree, never ripen and turn colour at the same time, so a maturity index has been developed to judge fruits maturity level. A maturity, index number helps to determine when each variety should be harvested in order to obtain the oil style the producer wants. The maturity index depends on the colour of the skin and flesh to assess maturity. A maturity index of 2.5 to 4.5 is usually used for most olive oils. At a maturity of 3.0 to 5.0

olives have reached their maximum oil content. The olives in a grove may reach this maturity index sooner or later in the year depending on weather but olives picked year after year at the same maturity index should produce similarly flavoured oil.

Maturity Index Formula

The maturity index can be determined on 100 randomly selected olives in each sample to obtain a numerical value for the olive sample appearance. Cut olives in half to expose the internal flesh and to permit grading (Figure 11).

The olives are categories using the following parameters:

0 = skin is a deep or dark green colour.

1 = skin is a yellow or yellowish-green colour.

2 = skin is a yellowish colour with reddish spots.

3 = skin is a reddish or light violet colour.

4 = skin is black and the flesh is completely green.

5 = skin is black and the flesh is a violet colour halfway through.

6 = skin is black and the flesh is a violet colour almost through to the stone.

7 = skin is black and the flesh is completely dark.

The total number of olives in each category are counted and recorded. The following equation is then applied to determine the maturity index:

$$\text{Maturity Index} = \frac{(0 \times no) + (1 \times n2)\ldots + (7 \times n7)}{100}$$

Where 'n' is the number of fruit with that score (Boskou, 1996).

This formula establishes the degree of maturity or maturity index of the sample. Further analyses are needed to determine the percentage moisture content; partial content of oil obtained by centrifugation on a fresh matter (f/m) basis; total oil content on a fresh matter (f/m) and dry matter (d/m) basis when obtained by solvent extraction, oil/solvent density or physical methods; extraction capacity; free acidity and sensory characteristics of the oil.

Determination of Maturity Indices for High Oil Recovery in Olive Under Temperate Condition (CITH, Srinagar)

Research study carried out at CITH, Srinagar to determine maturity indices for getting maximum oil extraction from the fruits was recorded and preliminary results showed that from second week of Oct. to first fortnight of Nov. month, was found be the best to harvest fruits to get higher oil percentage in 'Coratina', 'Leccino' and 'Pendolino'; however, in 'Messenese', 'Cipressino' and 'Picholine' from second fortnight of Oct. to end of Oct. found suitable for harvesting the fruits to get maximum oil yield.

Figure 11: Olive Fruit at Various Stages of Ripeness Based on the Maturity Index. It can be seen by M3 to M5 that black skin is not a good indicator that fruit have matured. (MI = maturity index). Adopted from Mailer *et al.*, 2005.

Fruit Harvesting Methods

Traditional Method

The traditional method of olive picking involves combing the ripe fruit from the tree into nets, or hand picking into baskets tied around the waist. Ladders are used to climb up into the trees to reach the fruit.

Modern Method

Mechanical pickers have in most places replaced hand combing technique, although there are increasing degrees of mechanisation of the olive harvest. At its most low tech pickers like a long handled vibrating tongs are used to remove the olives from the branches. The olives are collected in nets which have to be spread underneath the trees by hand. The next stage in automation relies on a shaker bar

Figure 12: Optimum Fruit Harvesting Stage for Higher Oil Recovery in different Varieties of Olive.

fitted to the back of a tractor. This shakes the tree from the trunk and prior to this unfurls a net around the base of the tree to collect the olives.

In order to produce an olive oil that is fresh, robust and valuable, it's important not only to harvest at right period, but to have an efficient picking process to get the olives quickly to the olive processing plant for pressing.

The Tools Required for Harvesting

Ladders

The single most important tool for non-industrial harvesting is a tall ladder. A leaning ladder may be used for inner branches and a step ladder for outer tree branches. Both are commonly used for hand harvesting and rake harvesting of olives. Olive trees can range in height from just a few feet to more than 20 feet tall unless pruned for size, so it's important to have at least one small 5- to 7-foot step ladder and another leaning ladder that can go as high as 25 feet. Pruning the trees as they grow can make reaching branches with a ladder much easier by reducing height.

Hands and Bags

The oldest and most labour-intensive olive harvesting method involves using hands and a large sack. Olive farmers pick the olives individually or in bunch with their hands and drop them into bags. Hand harvesting is time consuming and it can be difficult to reach the very highest olive branches. However, olives picked this way are much less prone to damage, are free of leaves or branch pieces and it's easier to select the best olives from the tree. This method is used most commonly for table olives, which need to be kept damage free. Yields per olive tree can vary enormously depending on tree size, but an average range is between 9 and 45 kg per harvest.

Olive Rake

The olive rake provides a step up in speed when compared to hand picking. A harvester uses the rake by laying a net below the olive trees and then running this specially designed rake through the branches, scraping and knocking the olives out and to the ground. Olive rakes are designed with elongated blunt tines to collect as many olives as possible without damaging them.

Netting

Another olive harvesting tool used by backyard and artisan growers is the olive net. These nets are usually made from non-toxic materials and either laid on the ground or arranged on a metal frame below the tree. The olive net is used to catch olives picked by hand, raked down from the branches, or shaken from the tree's branches, also by hand. A good netting material is nylon since its elasticity keeps olives from bruising.

Electric Olive Harvester

Although rakes and hands are the tools that have been used longest to harvest olives, some more modern hand tools have come on the market. One in particular is an olive harvesting electric whipper wand that uses a circle of moving tongs to knock ripe olives down from a tree's branches. The device attaches to a 7-foot pole and runs off a 12-volt battery. Electric wands like this reduce a lot of the labour required to pick olives manually.

References

CITH Annual Report. 2011. Central Institute of Temperate Horticulture Annual Report 2011-2012. Published by Director, CITH, Srinagar. pp. 79.

IOC. 2011. Guide for the determination of the characteristics of oil-olives. International Olive Council, COI/OH/Doc. No 1. pp. 2.

Mailer, R., Conlan, D. and Ayton, J. 2005. Olive harvest-harvest timing for optimal olive oil quality. RIRDC Publication No 05/013, RIRDC Project No DAN-197A. pp. 6-7.

19
Processing

The olive fruit is a drupe unlike peaches, apricots, and cherries *etc.*, but olives cannot be eaten straight from the tree as we can eat other drupe fruits. The olive contains a distinctive, bitter principle called *Oleuropein*. Oleuropein has a strong, bitter taste and although it is harmless, it is necessary to removed from olive before it can be consumed. Olives also have a low sugar content (2.6 – 6 per cent) compared with other drupes (12 per cent) and high oil content (20-30 per cent) depending on the time of year and cultivar. Primarily olives are processed for removal the bitterness, preservation of the fruit and enhancing their flavour.

The best way to process table olives is through a long, slow, natural fermentation in a brine solution. This process ensures that the olives maintain most of their natural goodness and rich olive taste. Green olives receive a lye treatment, before it is placed in large tanks with a brine solution. Black olives are simply rinsed and then placed in the tanks with brine. This process asks for careful monitoring of various factors throughout all the stages. It can take up to 9 months before the fruit are ready for packing. There is however other, easier and quicker methods which you might find more suitable to use at home. Unfortunately, with these methods most of the natural goodness and taste of the fruit is lost. For this reason it is necessary to add flavour in the form of vinegar and/or herbs like garlic *etc.* In India fresh olives are available from October to December. All olives are green initially and turns black as it ripens. Choose the right cultivar. Some cultivars are best processed when still green, like 'Manzanilla', 'Coratina' and 'Sevillano'. 'Belice', 'Lecinno', 'Pendolino', 'Kalamata and 'Mission' are best processed when black. For processing, the fruit must be fresh. Make sure the fruits have been stored in a cool place in clean, dry, well ventilated crates after harvest. The fruit must be firm to the touch, not shrivelled and free of any marks due to insect bites or stings. Table olives must be handpicked and handled with care to prevent bruising. Green olives must be full sized; the colour must be a slightly yellowish green. Black olives must be fully ripe, but still firm, the colour violet to violet-black. Hygiene is very important to prevent the olives from spoiling. So keep all the equipment (buckets, spoons *etc.*) clean. Also make sure the water you are using is clean, chlorine free and of a high quality. The various methods have been developed for the processing of olive.

Olive Processing Method

Long Term Method

Long term processing methods mainly rely to the natural fermentation process. It allows the sugars in the olives to ferment and form lactic acid. Using these methods, majority of the flavours are preserved along with the delicious taste. Green and black olives needs special treatments for processing.

Processing of Green Olives

Green olives are processed through a series of activities. Harvested fruits are to soak in a caustic soda solution (3-6 per cent) for ±15 hours. The time may vary according to the size and ripeness of the fruit. After a few hours, the soaked fruits are take out and cut in to equal halves. When the lye penetrates two thirds of the distance between the surface of the fruit and the pit, they are ready to be washed. Avoid olives to come in contact with air, as this darkens the colour or an unattractive khaki green. Keep in an airtight container (stainless steel, glass or high grade plastic will not affect the taste) through the entire process. In the mean time prepare the brine by dissolving 1 kg of salt in 10 litres of clean water. Slowly rinse the olives several times with clean, cold water to remove soapiness and caustic residue. Place the olives into a suitable container and cover completely with the brine. Make sure the container has a tight fitting lid. Leave to ferment for ±12 months. Taste them from time to time and decide according to taste. The bottling is done by removal of olives from the brine followed by rinsing with clean water and place into glass jars and cover with hot brine, covering and leave to cool. Store in a cool place and refrigerate after opening. Wine vinegar may be added to taste. Brine solution is prepared by dissolving 20 g of salt into 1 litre boiling water.

Processing of Black Olives

Black olives are processed similarly as green olives. In this process black olives are washed and left in fresh water for 1 or 2 days. The brine solution is prepared by dissolving 1 kg of salt in 10 litres of clean water. Place the olives into a suitable container and cover completely with the brine. Make sure that the container should be air tight. They are left for fermentation for ±12 months. The product is tasted from time to time till they are to your taste. Remove any olives that are floating on the surface. Cover immediately and leave to cool. Store in a cool place and refrigerate after opening.

Short Term Method

Processing of Green Olives

For short term processing green olives are soaked in a caustic soda solution of between 3-6 per cent for ±15 hours. The time may vary according to the size and ripeness of the fruit. After a few hours, take out an olive and make a cut through the flesh. When the lye penetrated two thirds of the distance between the surface of the fruit and the pit, they are ready to be washed. Also try to prevent the olives from coming into contact with air, as this can cause the colour to go dark or an unattractive khaki green. Keep in an airtight container (stainless steel, glass or high grade plastic will not affect the taste) through the entire process. In the mean time prepare the brine by dissolving 1 kilogram of salt in 10 litre of clean water. Now rinse the olives many times with clean, cold water to remove soapiness and caustic residue. This step is very important, because you don't want your olives to taste of caustic soda or "soapy". Now place the olives in a suitable container and cover completely with fresh water. Change the water twice a day for a minimum of 4 to 6 weeks. The more bitter you like your olives, the shorter the soaking period. In an airtight container, cover the olives with the brine and allow soaking for 4 months or longer, depending on personal taste. Bottling is made by placing olives into glass jars. Cover with hot brine: 20 g Salt mixed into 1 litre boiling water. Add a film of good quality Extra Virgin Olive Oil on top of the brine for better quality. Cover immediately and leave to cool. Store in a cool place and refrigerate after opening. Prepare a brine solution: 750 g to 1 kg Salt, 10 litre of water, Wine vinegar to taste (± 1 litre).

Processing of Black Olives

Black olives are processed by washing the olives to remove dust and dirt. Make a cut lengthwise through the flesh. Now place the olives in an airtight container and cover completely with fresh water. Change the water twice a day for a minimum of 4 to 6 weeks. The more bitter you like your olives, the shorter the soaking period. Prepare a brine solution: 750 g to 1 kg Salt, 10 litre of water, Wine vinegar to taste (± 1 litre). In an airtight container, cover the olives with the brine and allow soaking for 4 months or longer, depending on personal taste. Bottling is done by placing olives into glass jars. Cover with hot brine: 20 g Salt mixed into 1 litre boiling water. Add a film of good quality Extra Virgin Olive Oil on top of the brine for better quality. Cover immediately and leave to cool. Store in a cool place and refrigerate after opening.

Oil Characteristics of Varieties

The varietal character of oil is just like the varietal character of any fruit. It changes with the genetics of the variety. One of the most prominent components of the variety is the fatty acid composition. The fatty acids, sterols, methyl-sterols, and some alcohols are non-volatile compounds that do not add to the flavour of olive oil, but can influence the fluidity of the oil or mouth feel, as well as the stability of the oil and health aspects related to the amount of saturated versus mono and poly unsaturated fatty acids it contains. These components are also used to determine the authenticity or genuineness of oil. Another very important aspect of variety as it relates to oil flavour is the specific composition and quantity of the polyphenols and aromatic compounds it contains. The watery portion of the cell surrounding the globules of oil contains all the water-soluble and semi-water-soluble compounds, such as the polyphenols, tocopherols, glucosides, aldehydes, ketones, esters, organic acids, aromatic hydrocarbons, and pigments like chlorophyll and the carotenoids. The polyphenols and glucosides give the oil most of its bitterness, pungency, and together with the tocopherols, its antioxidant capacity. The volatile aromatic compounds such as the aldehydes, ketones, esters, and organic acids, plus some alcohols are responsible for much of the ultimate flavour of the oil. These are different than the polyphenols; they are not usually bitter or pungent, but give the oil certain characteristics sometimes described as flowery, ripe fruity, perfumey, *etc.* The pigment, chlorophyll, has some flavour, but mostly just gives olive oil a green colour. The variety essentially determines the quality of the fruit and oil. It has been well established that with distinct varieties distinct oil types are produced. The quantity of polyphenols, aromatic compounds, and many of these other compounds varies by variety, fruit maturity, and processing technique, which why there are so many different kinds of olive oil, each with its unique inherent flavour characteristics and stability. Some of the salient characteristics as described by Vossan (2007) are given below:

- ☆ *Arbequina*: Recognized for its aromatic ripe fruitiness, low bitterness, pungency and stability.
- ☆ *Aglandau*: Highly fruity, bitter, pungent and stable.
- ☆ *Barnea*: Fruity with mild bitterness, pungency and stability.
- ☆ *Bosana*: Highly fruity, herbaceous, medium pungency, bitterness and stability.
- ☆ *Chemlali:* Strongly aromatic, fruitiness with notable varietal character.
- ☆ *Coratina*: Strongly green, herbaceous, bitter, pungent and stable.
- ☆ *Cornicabra*: Very fruity and aromatic with medium bitterness, pungency and stability.
- ☆ *Empeltre*: Mildly fruity with low bitterness, pungency and stability.
- ☆ *Frantoio*: Very fruity, aromatic, and herbaceous; medium bitterness and stability; strongly pungent.
- ☆ *Hojiblanca*: Fruity, aromatic, mildly pungent, low bitterness and stability.
- ☆ *Koroneiki*: Strongly fruity, herbaceous, and very stable; mild bitterness and pungency.
- ☆ *Lechin de Sevilla*: Very fruity, mildly bitter, pungent and stable.
- ☆ *Leccino*: Medium fruitiness, and stability; low bitterness and pungency.
- ☆ *Manzanillo*: Fruity, aromatic and herbaceous; medium bitterness and stability and strongly pungent.
- ☆ *Moraiolo*: Very strongly fruity, herbaceous, and stable; medium bitterness and pungency.
- ☆ *Picudo*: Very aromatic, ripe fruitiness; medium pungency and stability and mildly bitter.
- ☆ *Picual*: Controversial variety that when harvested early produces a nicely aromatic fruity oil that has medium bitterness and very high stability. Poor reputation is due to poor fruit handling.

- ☆ *Picholine*: Very fruity and aromatic; medium fruitiness, bitterness and pungency.
- ☆ *Picholine Marocaine*: Very fruity and aromatic; medium fruitiness, bitterness and pungency.
- ☆ *Taggiasca*: Mildly fruity; low bitterness, pungency and stability.
- ☆ *Verdial de Huevar*: Mildly fruity, bitter, and pungent; very green in colour.

Refrence

Vossen, P. 2007. Olive oil: history, production, and characteristics of the World's classic oils. *Hortscience* **42**(5) : 1093-1100.

20

Value Added Products

Processing and value addition are the important aspects for olive fruits. The fruits are being used for various purposes and are being processed accordingly viz., oil, pickle and culinary *etc.* Olives are being cured since ancient times that make naturally bitter fruit into a deliciously salty, tart snack. Therefore, it is very much essential to choose a curing appropriate curing method that can works best for the most of the olive genotypes (green, purple or black). Water curing, brining, dry curing and lye curing each yield distinctly different flavors as well as textures. However, curing olives takes a long time but once cured, olives can be stored with flavorings (lemon, oregano, garlic, and others) and can use for different value added products preparation.

Methods of Olive Curing

1. Water Curing (Suitable for Large Green Olives)

In water curing process, fruits are subjected to wash properly followed by with stone or mallet, cracking of olive flesh and protecting them against any kind of bruise to the pit. This is usually done in a pan by dipping in cold water for 6-8 days. The water is needs to be changed twice a day in morning and evening, until the bitterness is removed (taste to test). The curing of olives is completed during this 6-8 days period. These are ready to fill in a pan with brine (about 1 part salt to 10 parts water) and lemon juice (about 1 part lemon juice to 10 parts water), if require transfer to jars and refrigerate for several hours before eating.

2. Brine Curing (Suitable for Black Olives)

In brine curing process, washing of the fruit is prerequisite to remove the dust and dirt of the field and transit. The water quality should be soft and clean. Thereafter, make a sharp cut in the pulp of the olive (top to bottom) without cutting the pit by using a clean and sharp knife. The brine curing is usually practiced in a pan by soak the olives in brine (1 part salt to 10 parts water). Make sure the olives are submerged (use something to weight them down) in the brine water and cover them properly. In this curing process usually more than 3 weeks period of time is required depending upon the varieties. Shake the pan each day as well as change the brine solution every week. Taste the bitterness (they could take up to 5-6 weeks depending on the olives) of the olives. When they taste the desirable standards, place in jars with brine (1 part salt to 10 parts water), add 4 tablespoons of red wine vinegar and top with a layer of olive oil.

3. Dry (Salt) Curing (Suitable for Large Black Olives)

The dry curing for large black olives are usually practiced outdoors, in a basket, burlap bag, or wooden box lined with burlap (that allows air to circulate), layer olives with coarse sea salt. Leave the olives outside (with plastic underneath to catch the juices that drain) for 3-4 weeks, shaking daily and adding a little more salt every 2-3 days. Taste for bitterness (rinsing the olive first). When no longer bitter, then shake off excess salt and keep them that way, or shake off the excess salt and dip them quickly in boiling water to get rid of the salt. They can be marinated for a few days in olive oil to regain plumpness (this type of curing will shrivel them), or just coated well with olive oil (using your hands) before eating.

4. Dry (Salt) Curing (Suitable for Small Black Olives)

The dry curing for small black olives is usually practiced in glass jars, alternate layers of olives with coarse salt. Every day for 3 weeks, shake well and add more salt to absorb the juices. Test the bitterness (rinsing the olive first). Continue to cure if bitterness remains, otherwise, add warm water to cover and 4 tablespoons of good quality red wine vinegar and top with a layer of olive oil. They are ready to eat after 4-5 days.

5. Lye Method

Lye also known as sodium hydroxide or caustic soda, is a multipurpose chemical that is commonly used in making soap, peeling canning types of peaches and in preparing some foods. Olive can rapidly be cured by placing them in a lye (sodium hydroxide) solution. The lye breaks the chemical bond between oleuropein (bitterness compound) and sugars in the olives. After curing is complete there must be removal of all traces of lye with a series of cold water rinses and then pack the olives in brine. The rinse process also removes the bitterness, leaving a neutral, somewhat' uttery' flavored olive that can accept flavours from vinegar and herbs in the brine. There is no fermentation step in this method. Lye cure olives have a firm texture and a smooth, mild taste. Lye cured olives can be stored for up to 2 month in brine, a described below or they can be preserved for longer storage by freezing, drying or pressure canning.

Value Added Product

1. Extra Virgin Olive Oil

According to the international standards, extra virgin is the highest quality and most expensive olive oil. It is free from any kind of defects and a unpleasant flavour. In chemical terms, extra virgin olive oil is described as having a free acidity, expressed as oleic acid, of not more than 0.8 grams per 100 grams and a peroxide value of less than 20 mill equivalent O_2. It must be produced entirely by mechanical means without the use of any solvents, and under temperatures that will not degrade the oil (less than 30°C). In order for an oil to qualify as "extra virgin" the oil must also pass both an official chemical test in a laboratory and a sensory evaluation by a trained tasting panel recognized by the International Olive Council. The olive oil must be found to be free from defects while exhibiting some fruitiness.

2. Olive Pickle

Olive pickles are very delicious and easy to make, but it takes very long time. There are many ways to prepare olives. Olives can either be picked early in the season while still green, or left to ripen and picked when they have just turned black. Both can be used for pickling, but the majority of black olives are pressed for oil. For preparation of olive pickle pick olives carefully, placing them into a bucket partially filled with clean water to avoid bruising the olives. Then make a salt solution large enough to cover the olives by dissolving 100 grams of coarse or cooking salt per 1 liter of water. Toss olives into the solution making sure they are completely submerged weigh down if necessary after that pour the solution away each day and replace with fresh solution and repeat this washing process for about 12 days for green olives or about 10 days for black. The best way to test when olives are ready is by the bite test. When the bitterness has nearly gone the olives are ready for the final salting then drain the solution and place olives in jars and measure the quantity of water required to cover olives in the jars and dissolve 100 grams of salt per 1 litre of water. Bring to boil and allow cooling. Pour salt water brine over the olives until completely submerged. Top up jars with up to 1 centimetre of olive oil to stop air getting to the fruit and seal the lids. No further preparation is required and the olives will store for at least 12 months in a cool cupboard. When they are ready to eat, pour off the strong brine and fill jar with clean water. Leave in fridge for a further 24 hours (the plain water leaches some of the salt back out of the olive). At this stage, add some flavoring agents like garlic, basil, oregano, chopped onion, red capsicum, lemon juice, and lemon pieces.

3. Olive Jam

Olive jam is another value added product and can be prepared by using ingredients such as drained and pitted 'Kalamata' olives, drained and pitted high-quality green olives, sugar, water, lemon – organic, apple – peeled, cored and diced and mild honey. For preparation of olive jam take large sauce pan, place all the olives (both kinds) and cover with cold water. Bring to a boil over medium-high heat and boil for one minute. Drain completely. Repeat this process two more times (this technique will remove enough salt from the olives so that they'll remain only mildly salty). Set the olives aside and rinse out the sauce pan. Add the sugar and the water to the sauce pan and swirl to combine the sugar with the water. Cut two or three of strips of zest (the colored part of the peel) from the lemon and drop the strips into the sugar water. Slice the lemon very thinly and add the slices to the pan. Bring this mixture to a simmer over medium heat and allow it to simmer for about 8-10 minutes, or until it's reduced to about one cup of liquid. Pour the liquid through a strainer into a bowl, pressing on the lemon solids to extract as much liquid as possible. Return the liquid to the pan, and add the diced apple, honey and the reserved olives. Bring the mixture to a simmer once again and simmer, stirring occasionally, just until the apples are soft and the liquid has become very thick – about another 10 to 15 minutes (add a bit of water if the mixture seems to be getting too thick before the apples are soft). Remove the pan from the heat and allow the mixture to cool slightly. Using an immersion blender (or using a regular blender), process the entire mixture until it is velvety-smooth. It should be a thick, "jammy" consistency already; if it seems runny, simply continue to cook it a little bit longer (stirring constantly), but keep in mind it will thicken as it cools. Transfer to jars and refrigerate.

4. Olive *Chatni*

It is made from Indian olives which are also called *'Jolpai'*. For this product, only Indian olive is suitable. These are typically bitter in taste. *Jalpai chutney* is a mouth watering sweet dish. This is thick chutney served after the meal. The ingredients required for making olive chutney are *'Jolpai'* or green olives, grated coconut, few raisins, half dried red chilli, fennel seeds, mustard seeds, turmeric powder, mustard oil and salt and sugar to taste. For preparation of olive chatni first boil the olives for about 2 whistles in pressure cooker. Add two glasses of water. Take out the olives and bring it to room temperature. Lightly smash the olives. Heat mustard oil then add dried red chilli, fennel seeds and mustard seeds to sputter in oil. Now add the smashed olives and cook it for a couple of minutes. Add salt and turmeric powder and mix it well. Add the water from the boiled 'Jolpai'. Add sugar and thereafter add soaked raisin and grated coconut. Cook it for 5 more minutes on medium heat. Remove from flame when ready.

Other Value Added Products

1. Olive syrup
2. Olive vinegar
3. Olive tea
4. Olive sweets
5. Crude olive cake- the residue of the first extraction of oil from the whole olive by pressure. Its relatively high water (24 per cent) and oil (9 per cent) content cause rapid spoilage when it is exposed to air.
6. Exhausted olive cake- the residue obtained after extraction of the oil from the crude olive cake by a solvent, usually hexane.

7. Partly destoned olive cake- the result of partly separating the stone from the pulp by screening or ventilation: it is called "fatty" if the oil has not been solvent-extracted. It is called "exhausted" or "defatted" if the oil has been solvent-extracted.
8. Olive pulp- the paste obtained when the stone has been separated from the pulp before extraction of the oil. It has high water content (60 per cent) and is difficult to store.
9. Vegetation waters- the brown watery liquid residue which has been separated from the oil by centrifugation or sedimentation after pressing.
10. Leaves collected at the oil mill- these are not pruning residues, but the leaves obtained after the olives have been washed and cleaned on entering the oil mill.
11. Body lotion: enjoy the skin nourishing properties of olive oil and organic essential oils in this SPF 15 body lotion with a high olive oil content and sweet orange aroma.
12. Lip balm: This natural lip balm comes in a tea tree-peppermint scent. The balm includes olive oil and organic cocoa butter, with SPF 15 protection.
13. Luxury soap: the olive oil luxury soap comes in three varieties: lemongrass-rosemary, fresh olive and sweet grass. Each hand-crafted bar is elegantly wrapped in burlap and contains natural ingredients such as olive oil, organic coconut oil, organic palm oil, organic essential oils and Vitamin E.

21
Quality Standards

There are mainly three different types of standards has been formulated by Codex Alimentarius for olive based products like table olive, olive oil and pomace oil.

Standards for Table Olives

These standards applies to the fruit of the cultivated olive tree (*Olea europaea* L.), which has been suitably treated or processed, and which is offered for direct consumption as table olives, including for catering purposes or olives packed in bulk containers which are intended for repacking into consumer size containers. For trade purposes olives have been into three different categories according to the degree of ripeness of the fresh fruits:

- ☆ *Green olives*: These table olives fruits must be harvested when they have reached normal size at ripening period, just prior to colouring.
- ☆ *Olives turning colour*: In this category, fruits should be harvested before the stage of complete ripeness is attained, at colour change.
- ☆ *Black olives*: In this category, fruits should be harvested at full ripening or slightly before full ripeness is reaches.

Generally, olives have been classified for trade purposes in to different categories based on the overall quality traits. There are three categories have been adopted based on certain standards by Codex Alimentarius in 2015.

i) *Extra or Fancy or A*: This category of fruits are of high quality with the characteristics specific to the variety. Fruits also notwithstanding, and providing this does not affect the overall traits or organoleptic characteristics of each fruit, they may have very slight colour, shape, flesh-firmness or skin defects. Whole, split, stoned (pitted) and stuffed olives of appropriate varieties may be classified in this category.

ii) *First, 1st, Choice or Select or B*: This category covers good quality olives with a suitable degree of ripeness and endowed with the characteristics specific to the variety and trade preparation. Providing this does not affect the overall favourable traits or individual organoleptic characteristics of each fruit, they may have slight colour, shape, skin or flesh-firmness defects. All the types, preparations and styles of table olives may be classified in this category, except for chopped or broken olives.

iii) *Second, 2nd or Standard or C*: This category includes good quality olives which, although they cannot be classified in the two previous categories, comply with the general conditions defined for table olives.

Standards for Olive Oils and Olive Pomace Oil (IOC, 1981)

The different designations or the types have been made for the olive oil. The olive oil is the oil obtained solely from the fruit of the olive tree (*Olea europaea* L.), to the exclusion of oils obtained using solvents or re-esterification processes and of any mixture with oils of other kinds. It is broadly classified in to two classes-

i. *Virgin olive oils*: These are the oils obtained from the fruit of the olive tree solely by mechanical or other physical means under conditions, particularly thermal conditions, that do not lead to alterations in the oil, and which have not undergone any treatment other than washing, decanting, centrifuging and filtration.

ii. *Olive-pomace oil:* This is the oil obtained by treating olive pomace with solvents other than halogenated solvents or by other physical treatments, to the exclusion of oils obtained by re-esterification processes and of any mixture with oils of other kinds.

Composition and Quality of Olive and Pomace Oils (IOC, 1981)

a. ***Extra virgin olive oil***: Virgin olive oil with a free acidity (≤0.8 g/100 g, as oleic acid).

b. ***Virgin olive oil***: Virgin olive oil with a free acidity (≤2.0 g/100 g, as oleic acid).

c. ***Ordinary virgin olive oil***: Virgin olive oil with a free acidity (≤3.3 g/100 g, as oleic acid).

d. ***Refined olive oil:*** Olive oil obtained from virgin olive oils by refining methods which do not lead to alterations in the initial glyceridic structure. It has a free acidity (≤0.3 g/100 g, as oleic acid).

e. ***Olive oil***: oil consisting of a blend of refined olive oil and virgin olive oils suitable for human consumption. It has a free acidity (≤1.0 g/100 g, as oleic acid).

f. ***Refined olive-pomace oil***: Oil obtained from crude olive-pomace oil by refining methods which do not lead to alterations in the initial glyceridic structure. It has a free acidity (≤0.3 g/100 g, as oleic acid).

g. ***Olive-pomace oil***: Oil consisting of a blend of refined olive-pomace oil and virgin olive oils. It has a free acidity, expressed as oleic acid (≤1.0 g/100 g, as oleic acid).

Table 8: Organoleptic Characteristics (Odour and taste) of Virgin Olive Oils

Category	*Median of the Defect*	*Median of the Fruity Attribute*
Extra virgin olive oil	Me = 0	Me > 0
Virgin olive oil	0 < Me ≤ 2.5	Me > 0
Ordinary virgin olive oil	2.5 < Me ≤ 6.0 *	

* Or when the median of the defect is less than or equal to 2.5 and the median of the fruity attribute is equal to 0. **(IOC, 1981)**

Table 9: Fatty Acid Composition of the Different Categories of Olive Oils as Determined by Gas Chromatography (in terms of per cent total fatty acids)

	Virgin olive oils	*Olive Oil Refined Olive Oil*	*Olive-Pomace Oil Refined Olive-Pomace Oil*
		Fatty acid	
C14:0	0.0–0.05	0.0–0.05	0.0–0.05
C16:0	7.5 – 20.0	7.5–20.0	7.5–20.0
C16:1	0.3–3.5	0.3–3.5	0.3–3.5
C17:0	0.0–0.3	0.0–0.3	0.0–0.3
C17:1	0.0–0.3	0.0–0.3	0.0–0.3
C18:0	0.5–5.0	0.5–5.0	0.5–5.0
C18:1	55.0–83.0	55.0–83.0	55.0–83.0
C18:2	3.5 – 21.0	3.5–21.0	3.5–21.0
C18:3			
C20:0	0.0–0.6	0.0–0.6	0.0–0.6
C20:1	0.0–0.4	0.0–0.4	0.0–0.4
C22:0	0.0–0.2	0.0–0.2	0.0–0.3
C24:0	0.0–0.2	0.0–0.2	0.0–0.2
		Trans fatty acids	
C18:1 T	0.0–0.05	0.0–0.20	0.0–0.40
C18:2 T+C18:3T	0.0–0.05	0.0–0.30	0.0–0.35

Table 10: Sterol and Triterpene Dialcohol Composition

Desmethylsterol Composition	*Per cent Total Sterols*
Cholesterol	≤ 0.5
Brassicasterol	0.2 for olive-pomace oils
	≤ 0.1 for other grades
Campesterol	≤ 4.0
Stigmasterol	≤ campesterol
Delta-7-stigmastenol	≤ 0.5
Beta-sitosterol + delta-5-avenasterol + delta-5-23-stigmasta-dienol + clerosterol + sitostanol + delta-5-24-stigmastadienol	≤ 93.0

Total Sterols Counts

Virgin olive oils Refined olive oil Olive oil	≥1,000 mg/kg
Crude oil-pomace oil	≥2500 mg/kg
Refined olive-pomace oil	≥1,800 mg/kg
Olive-pomace oil	≥1600 mg/kg

Erythrodiol and Uvaol Content (Per cent total sterols)

Virgin olive oils Refined olive oil Olive oil	≤ 4.5

Table 11: Wax Content

Virgin olive oils	≤ 250 mg/kg
Refined olive oil	≤ 350 mg/kg
Olive oil	≤ 350 mg/kg
Refined olive-pomace oil	> 350 mg/kg
Olive-pomace oil	> 350 mg/kg

Table 12: Maximum difference between the Actual and Theoretical ECN 42 Triglyceride C

Virgin olive oils	0.2
Refined olive oil	0.3
Olive oil	0.3
Olive-pomace oils	0.5

Maximum stigmastadiene content

Virgin olive oils = 0.15 mg/kg

Table 13: Peroxide Value

Virgin olive oils	<20 milliequivalents of active oxygen/kg oil
Refined olive oil	<5 milliequivalents of active oxygen/kg oil
Olive oil	<15 milliequivalents of active oxygen/kg oil
Refined olive-pomace oil	< 5 milliequivalents of active oxygen/kg oil
Olive-pomace oil	< 15 milliequivalents of active oxygen/kg oil

Table 14: Absorbency in Ultra-Violet K270

Category	*Absorbency in Ultra-violet at 270 nm*	*Delta K*
Extra virgin olive oil	≤ 0.22	≤ 0.01
Virgin olive oil	≤ 0.25	≤ 0.01
Ordinary virgin olive oil	≤ 0.30 (*)	≤ 0.01
Refined olive oil	≤ 1.10	≤ 0.16
Olive oil	≤ 0.90	≤ 0.15
Refined olive-pomace oil	≤ 2.00	≤ 0.20
Olive-pomace oil	≤ 1.70	≤ 0.18

* After passage of the sample through activated alumina, absorbency at 270 nm shall be equal to or less than 0.11. (IOC, 1981).

Food Additives

Virgin Olive Oils

No additives are permitted in these products.

Refined Olive Oil, Olive Oil, Refined Olive-Pomace Oil and Olive-Pomace Oil

The addition of alpha-tocopherols (d-*alpha* tocopherol; mixed tocopherol concentrate; dl-*alpha*-tocopherol to the above products is permitted to restore natural tocopherol lost in the refining process. The concentration of alpha-tocopherol in the final product shall not exceed 200 mg/kg.

Contaminants

Halogenated Solvents

Maximum content of each halogenated solvent = 0.1 mg/kg

Maximum content of the sum of all halogenated solvents = 0.2 mg/kg

Labelling

The products shall be labelled in accordance with the *General Standard for Labelling of Prepackaged Foods* (CODEX STAN 1 – 1985).

Other Quality and Composition Factors

Table 15: Quality Characteristics

	Maximum Level
Moisture and volatile matter:	
Virgin olive oil	0.2 per cent
Refined olive oil	0.1 per cent
Olive oil	0.1 per cent
Refined olive-pomace oil	0.1 per cent
Olive-pomace oil	0.1 per cent

Contd...

Table 15–*Contd...*

	Maximum Level
Insoluble impurities:	
Virgin olive oil	0.1 per cent
Refined olive oil	0.05 per cent
Olive oil	0.05 per cent
Refined olive-pomace oil	0.05 per cent
Olive-pomace oil	0.05 per cent
Trace metals:	
Iron (Fe)	3 mg/kg
Copper (Cu)	0.1 mg/kg

(IOC, 1981)

Organoleptic Characteristics

Table 16: Virgin Olive Oils

Others			
	Odour	*Taste*	*Colour*
Refined olive oil	Acceptable	Acceptable	light yellow
Olive oil	good	good	light, yellow to green
Refined olive-pomace oil	acceptable	acceptable	light, yellow to brownish yellow
Olive-pomace oil	acceptable	acceptable	light, yellow to green
Appearance at 20°C for 24 hours:			
Refined olive oil, olive oil, refined olive-pomace oil, olive-pomace oil:			Limpid

Table 17: Composition Characteristics

Saturated fatty acids at the 2-position in the triglyceride (sum of palmitic and stearic acids)	*Maximum Level*
Virgin olive oil	1.5 per cent
Refined olive oil	1.8 per cent
Olive oil	1.8 per cent
Refined olive-pomace oil	2.2 per cent
Olive-pomace oil	2.2 per cent

Chemical and Physical Characteristics

Relative density (20°C/water at 20°C): 0.910-0.916

Table 18: Refractive Index (n_D^{20}):

	Maximum Level
Virgin olive oil	1.4677-1.4705
Refined olive oil	
Olive oil	1.4680-1.4707
Olive-pomace oils	

Table 19: Saponification Value (mg KOH/g oil)

	Maximum Level
Virgin olive oil	184-196
Refined olive oil	
Olive oil	182-193
Olive-pomace oil	

Table 20: Iodine Value (Wijs)

	Maximum Level
Virgin olive oil	75-94
Refined olive oil	
Olive oil	
Olive-pomace oil	75-92

Table 21: Unsaponifiable Matter

	Maximum Level
Virgin olive oil	15 g/kg
Refined olive oil	
Olive oil	
Olive-pomace oil	30 g/kg

Table 22: Absorbency in Ultra-violet K232

	Absorbency in ultra-violet at 232 nm
Extra virgin olive oil	≤ 2.50[1]
Virgin olive oil	≤ 2.60[4]

Reference

IOC, 1981. International Olive Council, Codex standard for olive oils and olive pomace oils, *CODEX STAN* **33**: 1-9.

22
Designations and Definitions of Olive Oils

Olive oil is obtained from the storage organs (fruit) of olive tree. The oil accumulation in storage organs follows different patterns depending on the organ kinetic development. In olive, the kinetic is much slow as judged by the expression pattern of the stearoyl-ACP desaturase in the seed and the pulp up to 154 and 196 DAFB, respectively. In the olive drupe, the full bloom occurred by the second week of May and oil accumulation starts by the beginning of July (5 to 7 weeks after full bloom) and ends by October 22 to 25. The final ripening stage corresponds to water and fresh weight losses for fruits and this increases the final mill oil yield. In fruit oil such as for the olive, oil accumulates both in the seed and in the pulp (fruit oil), but the pathways are disconnected. They may have different composition for fatty acids. All the drupe part of fruits from one tree has the same genetic composition whereas the seeds have each another genetic composition resulting from the mother tree allele segregation for the female gamete and the alleles from the fertilizing pollen grain. The ratio oil from seed on oil from the pulp is weak and oil composition is considered as being due to pulp. Olive oil quality depends on several factors of production and processing operations performed in olive fruits. The quality of olive oil depends on the following production factors and they contribute quality by 5 to 30 percent. Percentage of quality as affected by factors responsible for production are-

Varietal character	-	20 per cent
Ripening stage	-	30 per cent
Harvesting methods	-	5 per cent
Transport facilities	-	5 per cent
Storage and their methods	-	10 per cent

But, the quality of olive oil from processing points depends on extraction methods (heat and cold methods) as much as by 30 per cent. Compared to cold method the heat method produces low quality oils.

Olive oils are graded by production method, acidity content, and flavour. The International Olive Oil Council (IOOC, 2015) sets quality standards that most olive-oil-producing countries use, but the United States does not legally recognize these benchmarks. Instead, the U.S. Department of Agriculture uses a different system that was set up before the IOOC existed. However, American olive growers and oil importers are encouraging the USDA to adopt standards similar to those of the IOOC. It is marketed in accordance with the following designations and definitions:

Virgin olive oils are the oils obtained from the fruit of the olive tree (*Olea europaea* L.) solely by mechanical or other physical means under conditions, particularly thermal conditions, that do not lead to alterations in the oil, and which have not undergone any treatment other than washing, decantation, centrifugation and filtration. Virgin olive oils fit for consumption and based on standards of IOC. These are categorised as below:

- ✰ **Ultra-premium extra virgin olive oil**: Free fatty acids cannot exceed 0.3 per cent (0.3 grams per 100 grams, expressed as oleic acid), the oleic acid monounsaturated fat content must be greater than 65 per cent (IOC standard is 55 per cent), the peroxide value must not exceed 9 mEQ O2/kg oil (IOC standard is 20), lower K-values which measure UV absorption rates, the "Fresh pack" test which measures DAGs (diacilglycerides) and PPPs (pyropheophytins). All UP oils will also have a minimum Polyphenol measured content (at time of crush) of not less than 130 ppm (mg caffeic acid/kg).
- ✰ **Extra virgin olive oil:** Virgin olive oil has a free acidity, expressed as oleic acid, of not more than 0.8 grams per 100 grams, and the other characteristics of which correspond to those fixed for this category in the IOC standard.
- ✰ **Virgin olive oil:** Virgin olive oil which has a free acidity, expressed as oleic acid, of not more than 2 grams per 100 grams and the other characteristics of which correspond to those fixed for this category in the IOC standard.
- ✰ **Ordinary virgin olive oil:** Virgin olive oil which has a free acidity, expressed as oleic acid, of not more than 3.3 g/100 g and the other characteristics of which correspond to those fixed for this category in the IOC standard. This designation may only be sold direct to the consumer if permitted in the country of retail sale. If not permitted, the designation of this product has to comply with the legal provisions of the country concerned.
- ✰ ***Virgin olive oil not fit for consumption*** as it is, designated *lampante virgin olive oil*, is virgin olive oil which has a free acidity, expressed as oleic acid, of more than 3.3 g/100 g and/or the organoleptic characteristics and other characteristics of which correspond to those fixed for this category in the IOC standard. It is intended for refining or for technical use.

- ☆ ***Refined olive oil*** is the olive oil obtained from virgin olive oils by refining methods which do not lead to alterations in the initial glyceridic structure. It has a free acidity, expressed as oleic acid, of not more than 0.3 g/100 g and its other characteristics correspond to those fixed for this category in the IOC standard. This designation may only be sold direct to the consumer if permitted in the country of retail sale.
- ☆ ***Olive oil*** is the oil consisting of a blend of refined olive oil and virgin olive oils fit for consumption as they are. It has a free acidity, expressed as oleic acid, of not more than 1 g/100 g and its other characteristics correspond to those fixed for this category in the IOC standard. The country of retail sale may require a more specific designation.
- ☆ ***Olive pomace oil*** is the oil obtained by treating olive pomace with solvents or other physical treatments, to the exclusion of oils obtained by re esterification processes and of any mixture with oils of other kinds. It is marketed in accordance with the following designations and definitions-
    - ❒ ***Crude olive pomace*** oil is olive pomace oil whose characteristics correspond to those fixed for this category in the IOC standard. It is intended for refining for use for human consumption, or it is intended for technical use.
    - ❒ ***Refined olive pomace oil*** is obtained from crude olive pomace oil by refining methods which do not lead to alterations in the initial glyceridic structure. It has a free acidity, expressed as oleic acid, of not more than 0.3 g/100 g and its other characteristics correspond to those fixed for this category in the IOC standard. This product may only be sold direct to the consumer if permitted in the country of retail sale.
    - ❒ ***Olive pomace oil*** is the oil comprising the blend of refined olive pomace oil and virgin olive oils fit for consumption as they are. It has a free acidity of not more than 1 g/100 g and its other characteristics correspond to those fixed for this category in the IOC standard. The country of retail sale may require a more specific designation.
    - ❒ ***Olive cake*** is the solid phase that's remained after pressing olives and also called pomace.

Other Classification of Olive Oil

- ☆ **Light olive oil**: "Light" Olive Oil is a marketing concept and not a true classification of Olive Oil grades. It is not a regulated designation, so there are no real parameters for what its content should be. Sometimes, the Olive Oil is blended with other vegetable oils. It is important to note that this designation refers to flavour only, not caloric content, as all types of Olive Oil have the same number of calories. This oil is often flavourless and of low quality. It is refined oil.
- ☆ **Blended olive oil:** This refers to the combining of olive oils from different groves, varieties, and qualities (sometimes from different countries also)

to create a blend that offers a desired taste. Changes in weather and other conditions impact the same olive variety in the same region differently every year. Since large supermarket brands must taste the same year over year, "master blenders" are employed to create a recipe that combines these different oils from different sources to create the same finished product consumers associate with a particular brand. Another reason for blending is to increase oil's shelf life. This is achieved by blending oil high in polyphenols with one that does not. Sometimes olive oil is blended with canola or other vegetable oils. This is legal only if stated on the label. Illegal blending of cheaper hazelnut oil can be profitable for the dishonest producer and is difficult to detect.

- ☆ **Organic olive oil:** Olive oil produced in a holistic, ecologically-balanced approach to farming, without the use of any pesticides or chemicals. It is important to note that many Olive Oils may be organically produced; however, the high cost of certification is often prohibitive for small independent producers. For this reason very few oils will actually be certified organic. Some of the more common certification agencies granting the organic designation include 'BIOHELLAS', 'ECOCERT', 'USDA' 'ORGANIC', 'DEMETER', 'ICEA', and 'DIO', although there are many more.
- ☆ **Unfiltered olive oil:** Unfiltered oil will contain small particles of olive flesh. While some claim this adds additional flavour, it often causes sediment to form at the bottom of the bottle. There are varying degrees of filtration (partially filtered, lightly filtered, *etc.*), but this sediment may become rancid. This will over time deteriorate the oil's flavour and shelf life. Unfiltered oil should be properly stored and used within 6 months of bottling. We recommend that unfiltered oils are best enjoyed drizzled over salads or grilled vegetables, and not for cooking.
- ☆ **Early harvest olive oil** or **fall harvest olive oil:** Olives reach their full size in the fall but may not fully ripen from green to black until late winter. Green olives have slightly less oil, more bitterness and can be higher in polyphenols. The oil tends to be more expensive because it takes more olives to make a bottle of oil. Many people like the peppery and bitter quality of early harvest oil. Flavour notes of grass, green, green leaf, pungent, astringent are used to describe early harvest fall oils. Because of the higher polyphenols and antioxidants, early harvest oils often have a longer shelf life and may be blended with late harvest oils to improve the shelf life of those late harvest oils.
- ☆ **Late harvest olive oil** or **winter harvest olive oil:** The fruit is picked black and ripe. The fruit may have a little more oil but it is risky because waiting longer into the winter increases the risk the fruit will be damaged by frost. Late harvest or "winter" fruit is riper so like other ripe fruit it has a light, mellow taste with little bitterness and more floral flavour. Flavour notes of peach, melon, perfumy, apple, banana, buttery, fruity, rotund, soave and sweet are often used.

- ☆ **Flavoured olive oil:** These olive oils of increasing popularity have been infused with herbs or fruits. Typically flavoured olive oils use a lower quality of olive oil so you must be cautious about these and always read the label.
- ☆ **Hand-picked:** This refers to olives that are literally individually picked by hand directly from the tree. The argument is that mechanical harvesting can bruise the fruit, which will result in a higher acidity, so to avoid this; olives are carefully picked by hand and are also referred to as "Hand Harvested".
- ☆ **Estate olive oil** or **single estate olive oil:** Oil labelled "estate olive oil" means that all olives were harvested, crushed, and bottled in the same olive grove. These oils are usually produced in small batches and are sought after by consumers for their uniqueness in flavour and profile. Occasionally the term "Single Estate" will be used instead and means the same thing. These oils tend to be more expensive and are of a higher quality.
- ☆ **Mono varietal olive oil:** Oil labelled "mono-varietal" or "single varietal" means that only one type of olive was used to produce that olive oil. These oils are often desired for the enjoyment of the most "pure" characteristics and flavour of an individual olive type.

Colour Considerations

Green olive oils come from unripe olives and impart a slightly bitter and pungent flavour. Emerald-tinged oils have fruity, grassy, and peppery flavours that dominate the foods in which you use them. These oils are great with neutral-flavoured foods that allow their bold flavours to shine by pairing green olive oils with strongly flavoured foods as long as they complement the oils' pungent tastes. Olive oils that glimmer with a golden colour are made from ripe olives. Olives turn from green to bluish-purple to black as they ripen. Oils made from ripe olives have a milder, smoother, somewhat buttery taste without bitterness. These oils are perfect for foods with subtle flavours because the gentle taste of a ripe olive oil won't overshadow mildly flavoured foods.

Olive Oil and Gastronomy

Nowadays the use of olive oil is no longer limited to areas where the olive tree is grown and it has come to represent quality cooking almost all over the world. The alteration undergone by vegetable oils when heated for frying is quicker and more intense the higher their content of polyunsaturated fatty acids (seed oils), and the higher the initial acidity of the oil (it is more stable if it has a high content of natural antioxidants - vitamin E). This alteration also varies according to temperature and length of time heated, number of times used, manner of frying (in continuous frying it changes less), and the type of food being fried (frying fish, especially oily fish, increases the polyunsaturated acid content of the oil, facilitating its decomposition). To retain its properties it must be kept away from excess heat, air, damp and above all, from light.

Designations

Some olive oils may also have additional designations that certify their growing area of origin. These areas or zones are determined by the European Union to promote and protect regional food products within the EU. This has helped to encourage diverse agricultural production, protect product names from misuse and imitation, and has helped consumers by giving them information concerning the specific character of the products. Similar systems and designations exist in the world of cheese, and wine, (*e.g.*, VQA wines in Canada). There are different designations have been existing are:

- ✰ ***PDO (Protected Denomination of Origin)***: The olive oil must be produced, processed and prepared in a specific region using the area's traditional production methods. For Spanish olive oils it is 'DO' or 'Denominación de Origin. For French olive oils is 'AOC' or 'Appellation d'OrigineContrôlée'.
- ✰ ***PGI (Protected Geographical Indication)***: This designation is slightly less stringent than PDO, but still requires that the product be produced in the specified geographical region. The geographical connection must exist for at least one stage of production, processing or preparation. If only one of the stages of production has taken place in the defined area, it still qualifies as PGI. As an example, the olives may come from another region. This allows for a more flexible connection to the region and can focus on a specific quality, reputation or other characteristic attributable to that geographical origin. In Italy, you may also see these oils labelled as IGP.
- ✰ ***TSG (Traditional Speciality Guaranteed)***: This designation highlights traditional character, either in the Olive Oil's composition or in its means of production.

Reference

IOOC. 2015. http: //www.internationaloliveoil.org/web/aa-ingles/oliveWorld/aceite.html.

23
Purity and Fidelity of Olive Oil

Olive oil is the oil extracted exclusively from fruit of *Olea europaea* L. only by means of mechanical methods or other physical procedures that do not cause any alteration of the glyceric structure of the oil thus preserving its characteristics and properties. The cost of the olive is very high as compared to other commonly used vegetable oils. Some cheaper olive oils are also comes up in the market which might be adulterated with cheaper oils. Blending premium olive oil with low quality oils (mostly pomace) or with other plant oils such as hazelnut (*Corylus avellana*), soya (*Glycine max*), almond (*Prunus dulcis*), maize (*Zea mays*), sunflower (*Helianthus annuus*) and sesame (*Sesamum indicum*) has a great negative effect on olive oil trade (IOC, 2013). Therefore, purity test reports are essential part of the olive oil trade. The standardized and recommended techniques for identification of olive oil are as under:

Olive Oil Purity Testing Methods

Chemical Methods

Several analytical techniques have been refined in recent past based on the better knowledge on olive oil composition and instrumentation based analytical techniques. Unfortunately, adulterations are too improved and became more sophisticated; for example, it is recognized that by careful blending high oleic oils (such as sunflower oil) the obtained product well fits the fatty acid composition of olive oil. However, they usually have a "wrong" *sterol* composition and those who perform frauds were able to eliminate sterols, because of this act, the absolute amount of sterols was enclosed in the standard as well as the measurement of sterol dehydration products (such as *stigmastadienes*). To relate the fatty acid composition of olive oils with the cultivar, researchers (Mannina *et al.*, 2003; Montealegre *et al.*, 2010) have studied olive oil in a well-limited geographical region, with no consideration of the pedoclimatic factor (soil characteristics such as temperature and humidity) and also a relationship between the fatty acid composition and some specific cultivars has been studied. The volatile fraction in olive oils, which represents one of the most important qualitative aspects of this oil, consists of a complex mixture of more than 100 compounds, but the most important substances useful for olive cultivar differentiation are the products of the lipoxygenase pathway (LOX). Only a subset of volatile compounds and a combination among them could provide valuable information for olive cultivar differentiation. In fact, genetic and geographic factors influence the volatile compound production of the olive fruits and affect the differentiation of olive oils according to their olive variety. The volatile compound contents allowed differentiation among monovarietal olive oils and even identification of the technique used for olive oil production. The colour of a virgin olive oil is due to the solubilisation of the lipophilic chlorophyll and carotenoid pigments present in the fruit. The green-yellowish colour is due to various pigments, that is, chlorophylls, pheophytins, and carotenoids. Several researchers reported the same qualitative composition in chlorophyll and carotenoid pigments, independent of the olive variety and the time of picking. The carotenoid and chlorophyll content determination using UV-VIS Spectrophotometry was not useful to discriminate oils produced from different olive varieties. Lutein/β-carotene ratio has been reported as a tool to differentiate oils from a single cultivar. Tocopherols and hydrocarbons are the compositional markers less studied to date to differentiate olive oils. An important common aspect is that the content and composition of these markers are highly affected by the environmental conditions. The International Olive Oil Council (IOC) approved the following tests:

i) Detection of Sterols Composition by Capillary-Column Gas Chromatography

In this technique the fatty substance is saponified with potassium hydroxide in ethanolic solution and the unsaponifiables are then extracted with ethyl ether. The á-cholestanol is used a as an internal standard. The sterol fraction is separated from the unsaponifiable extract by chromatography on a basic silica gel plate. The sterols recovered from the silica gel are transformed into trimethyl-silyl ethers and are analyzed by capillary-column gas chromatography.The sterol concentration is calculated in mg/kg of fatty material as follows:

$$\text{Sterol } x = A_x \times m_s \times 1000 / A_s \cdot m$$

where,

Ax = peak area for sterol x, in square millimetres;

As = area of the α-cholestanol peak, in square millimetres;

ms = mass of α-cholestanol added, in milligrams;

m = mass of the sample used for determination, in grams.

ii) Detection of the Presence of Extraneous Oils in Olive Oil by HPLC

This method is widely used around the world. In this, triacylglycerol (TAG) composition is compared with the theoretical obtained from the analysis of fatty acid methyl esters (FAME). The oil is purified by solid phase extraction (SPE) on silica gel cartridges. The TAG composition is determined by reverse phase high-resolution liquid chromatography (RP-HPLC) using a refractive index detector and propionitrile as the mobile phase. FAME is prepared from purified oil by methylation with a cold solution of KOH in methanol and then the esters are analysed by capillary gas chromatography using high polar columns. The theoretical TAG composition is calculated from the fatty acid composition by a computer program assuming a 1,3-random, 2-random distribution of fatty acids in the triacylglycerol, with restrictions for saturated fatty acids in the 2-position.

Molecular Markers (DNA) vs. Biochemical as Detection Tool

Olive oils labelled with their region of origin are sold at a premium price. This premium is greatest for oil from those regions associated with superior taste, consistency or colour. For cold-pressed oils (extra virgin and virgin), these properties are associated with the cultivar and the environment. A lot of research has been carried out to assure the authenticity of olive oil through chemical analysis. However, several difficulties have been encountered in distinguishing olives and olive-oils from different cultivars because their characteristics are strongly influenced by environmental conditions. On the other side, accurate and rapid identification of cultivars is especially important to obtain a reliable label of origin. In such a case, some DNA-based technologies can help in revealing either the authenticity or the different origin of lots that have contributed to the olive oil. This action discourage from the adulteration with extraneous material of lower cost and value.

i) Molecular Tools

Molecular marker approaches are reliable, fast and low-cost and can facilitate the monitoring of the authenticity and identification of the cultivars involved in the production of olive oil, even if the sample has been subjected to some transformation and processing. Therefore, the use of this technique for the purpose of oil traceability increases the value of the product. The use of olive cultivars with low suitability for oil production reduces the quality of the olive oil. Moreover, the practice of adulterating olive oil by adding other, lower-cost vegetable oils is a frequent practice in some areas. For mixtures, chemical analysis is sufficient to confirm the authenticity of the product. However, when the objective is to identify olive oil of different cultivars, it is necessary to investigate the presence of genomic DNA from varietal contaminants. In this case, molecular markers can be used because they are able to detect genetically distinct samples with high accuracy. Different markers are used to track the identity of olive oils to certify the quality of the final product. Alba *et al.* (2009) used microsatellite markers to evaluate the quality and origin of Italian olive oils produced from 7 certified olive cultivars. They compared the DNA from leaves against the olive oil from the fruits of the same plant. Doveri *et al.* (2006) used microsatellite markers, SCAR and SNPs to verify the presence of DNA of the pollen from the donor plant in olive oil from the cultivar Leccino. Pasqualone *et al.* (2007) also evaluated the efficiency of microsatellite markers for the identification of Italian olive oil using different mixtures of samples from olive oils produced from

5 different olive cultivars. Despite the importance and applicability of molecular marker techniques in the olive processing industry, some studies have reported some concerns that could reduce their efficiency.

The concerns in molecular traceability detection:

1. The high-level degradation of genomic DNA during the process of oil production, which reduces the efficiency of DNA amplification and identification.
2. The availability of one efficient method for DNA extraction from olive oil is crucial for the incorporation of molecular marker techniques for routine quality analysis.
3. International standardization of sampling and analysis methods, as well as the creation of a network of reference laboratories, are also required to validate results.
4. RAPDs and AFLPs give complex profiles that can be applicable to monovarietal oils but not to mixtures of three to four cultivars, such as those usually adopted in PDO oils.
5. Single locus microsatellites are more effective to this aim, but they are not applicable to high throughput screening such as microarray.
6. Single locus (SSRs and SNPs) are preferred to multilocus (AFLP, RAPD) markers because they are simpler to perform, more easily interpretable and can be combined in high throughput platforms.
7. Since most olive cultivars are auto-incompatible (pollen could not germinate on an ovary from the same tree) the DNA extracted from oil contains alleles of the tree (fruit pulp somatic tissues) as well as alleles of the seed embryo which may contain exogenous alleles from the pollinator. Thus, care needs to be taken in the interpretation of DNA profiles obtained from DNA extracted from oil for resolving provenance and authenticity issues. Besides, using capillary electrophoresis permits to differentiate alleles with very small differences in molecular weight and to detect a very low or partially degraded DNA, which is the case of extracted DNA from olive oil. The availability of other approaches such as semi-automated SNP genotyping assay proposed to verify the origin and authenticity of monovarietal extra virgin olive oils.

The generation of an "Identity Card" which, can be used for the unequivocal identification of highly prized oil, has been considered as a potential for olive oil DNA fingerprinting.

Future Challenges and Prospective

Future research towards olive oil genetic traceability will concern the following aspects given below:

1. Application of high-throughput platforms including functional genes, non-nuclear genes, transcriptome analysis, and developing more sophisticated SNPs detection.

2. Understanding the function of genes and other parts of the genome is known as functional genomics that describes the relationship between an organism's genome and its phenotype. This approach provides a more complete picture of how biological function arises from the information encoded in an organism's genome and such information will contribute in intraspecies determination especially with PDO oils.
3. Chloroplast DNA considers as a most important non-nuclear genes and has been investigated for cultivar identification in olive oil. One advantage of chloroplast DNA is the high copy number of chloroplast per cell (about 50), which is especially beneficial for refined oil sample. It is to develop suitable markers on this region and a compositional test able to identify a cultivar in a monovarietal olive oil. The designed markers will be applied in a high-throughput platform to assess and quantify the contribution of a single cultivar in commercial multivariate oils.
4. Olive transcriptome will address the identification of genes differentially expressed during fruit development, with particular attention to those involved in lipid and phenolic metabolism. The provided information will discuss the case of olive oil PGI. Improving SNPs detection using high resolution melting (HRM) RT-PCR analysis allows olive cultivar genotyping, results in an informative, easy, and low-cost method able to greatly reduce the operating time is also recommended.
5. An alternative strategy would be using fast and less accurate sensor technology, such as electronic nose, as screening method and verifying suspected samples by DNA method.

References

Alba, V., Sabetta, W., Blanco, A., Pasqualone, A., Montemurro, C. 2009. Microsatellite markers to identify specific alleles in DNA extracted from monovarietal virgin olive oils. *Eur. Food Res. Technol.* **229**(3): 375-382.

Doveri, S., O'Sullivan, D.M., Lee, D. 2006. Non-concordance between genetic profiles of olive oil and fruit: a cautionary note to the use of DNA markers for provenance testing. *J. Agric. Food Chem.* **54**(24): 9221-9226.

IOC, 2013. Global method for the detection of extraneous oils in olive oils. IOC. http: //www.internationaloliveoil.org/

Mannina, L., Dugo, G., Salvo, F., Cicero, L., Ansanelli, G., Calcagni, C., Segre, A. 2003. Study of the cultivar-composition relationship in Sicilian olive oils by GC, NMR, and statistical methods. *J. Agric. Food Chem.* **51**: 120–127.

Montealegre, C., Alegre, MLM., Garcia-Ruiz, C. 2010. Traceability markers to the botanical origin in olive oils. *J. Agric. Food Chem.* **58**(1): 28–38.

Pasqualone, A., Montemurro, C., Summo, C., Sabetta, W., Caponio, F., Blanco, A. 2007. Effectiveness of microsatellite DNA markers in checking the identity of protected designation of origin extra virgin olive oil. *J. Agric. Food Chem.* **55**(10): 3857-3862.

24

Method and Sampling Techniques for Olive Oil Quality Analysis

Sampling as well as analytic techniques influence the test results. The International Olive Oil Council and Private and Public funded laboratories have been working for standardization of the most efficient as well as accurate methods based on the different components of the olive oil. The olive oil mainly composed of triacylglycerols (triglycerides or fats) and contains small quantities of free fatty acids (FFA), glycerol, phosphatides, pigments, flavour compounds, sterols, and microscopic bits of olive. Triacylglycerols are the major energy reserve for plants and animals. Chemically speaking, these are molecules derived from the natural esterification of three fatty acid molecules with a glycerol molecule. Based on USDA Nutrient Database, 100 g olive oil contains, 3699 kJ (884 kcal) energy, 100 g fat. In 100 g fat, 14 g is saturated, 73 g monounsaturated, 11 g polyunsaturated (0.8 g omega3 and 9.8 g omega6 types). It also contains 14 mg Vitamin, 60 μg Vitamin K and 0.56 mg Iron. Different methods have been recommended for estimation of oil content, moisture content, free fatty acid, acidity, peroxide value, specific absorption coefficient, anisidine value, iodine value, colour characteristics and sensory properties. The brief procedure given as below:

1. Determination of Oil Content

Soxhlet Extraction Method

The oil content is estimated by Soxhlet solvent extraction method. Solvent extraction removes all traces of oil from the olive material. The extracted oil does not meet the requirements of extra virgin oil, but it is the only way of determining the absolute amount of oil in the fruit. The method gives a result higher than you would expect to get from an olive press, but it provides an accurate comparison between trees.

Equipment and other Materials Required

Soxhlet apparatus, hot plate, balance, boiling stones, copper bowl, rubber tubes, thimble, glass rod, wood box, shredded cardboard, stand, round bottom flask, clamps, olive oil, anhydrous diethyl ether, 200-300 ml beaker for mixing oil and cardboard, and potato chips.

Procedure

Weigh a clean thimble before and after roughly the same amount of shredded cardboard is added as in the hexane/isopropanol experiment (160-180 g). Again add 1-4 g of olive oil and mix it with glass rod. Insert the thimble into the Soxhlet apparatus, and turned on hot plate for ether filled round bottom flask. It is important to note that boiling stones are added prior to any heat is applied to the ether. Once the solvent is start boiling at a steady rate, the equipment is left to "run" through seven refluxes. A reflux is when enough solvent is collected in the Soxhlet cylinder to be siphoned back into the round bottomed flask. After completion of it finished, the thimbles are allowed to dry in a fume hood overnight. They are then placed in an air-tight chamber to prevent water from being absorbed or lost until they are later weighed and the per cent fat extract calculation.

Cold Press Method

In this method the recovery of the oil is less as compared to Soxhlet extraction method. It is also difficult to compare the oil produced by two trees as the result varies slightly each time. However, it gives a better indication than solvent extraction of the oil that might be extracted by an olive press. In addition, the oil extracted is extra virgin oil and can be used for subsequent tests.

NIR (Near infra-red analyser) Method

This method can be used in the laboratory or processing mill. NIR technology can measure oil and moisture in milled olives and fatty acids and moisture in oil. The instrument works by passing near infra-red light through oil samples and measuring absorbed energy which relates to the concentration of oil, moisture and fatty acids.

2. Determination of the Moisture Content

Moisture content of olives changes dramatically from day to day. If the moisture content in the olive is high, the fruit will be heavy and the oil concentration will be diluted. High moisture content also means oil extraction may be restricted. There may also be a loss of flavour and reduced levels of antioxidants including polyphenols, although the oil content does not change. It is therefore important to test the moisture content of the olives at the same time that oil content is tested so that samples can be accurately compared with each other and over time. Fresh fruit moisture content is often 55 per cent or higher in developing fruit but then drops to 50 per cent or less with the commencement of ripening. This will vary between varieties and seasonal conditions and some varieties will not follow this trend. Cultivar 'Manzanillo' tend to maintain a high moisture level throughout the ripening period. By monitoring fruit moisture levels from the beginning of ripening period, growers can utilise moisture content to indicate optimum harvest timing and determine if it is conducive to good extraction and separation efficiency. Generally evaporation method is used for estimation of moisture content in the olive fruits. This method relies on measuring the mass of water in a known mass of sample. The moisture content is determined by measuring the mass of a olive before and after the water is removed by evaporation.

3. Determination of Free Fatty Acids

Determination of the free fatty acids (FFA) is the simplest test and indicative of good harvesting and handling processes. FFAs are fatty acids which have broken away from oil molecules or triacylglycerols. Their presence indicates that degradation has occurred in the oil through poor handling during processing. Free fatty acids can influence the organoleptic value of the oil. Additionally, free fatty acids are water-soluble and may be lost during processing. Every batch of oil should be tested for free fatty acids. The IOC standard for free fatty acids in extra virgin olive oil is a maximum of 0.8 per cent.

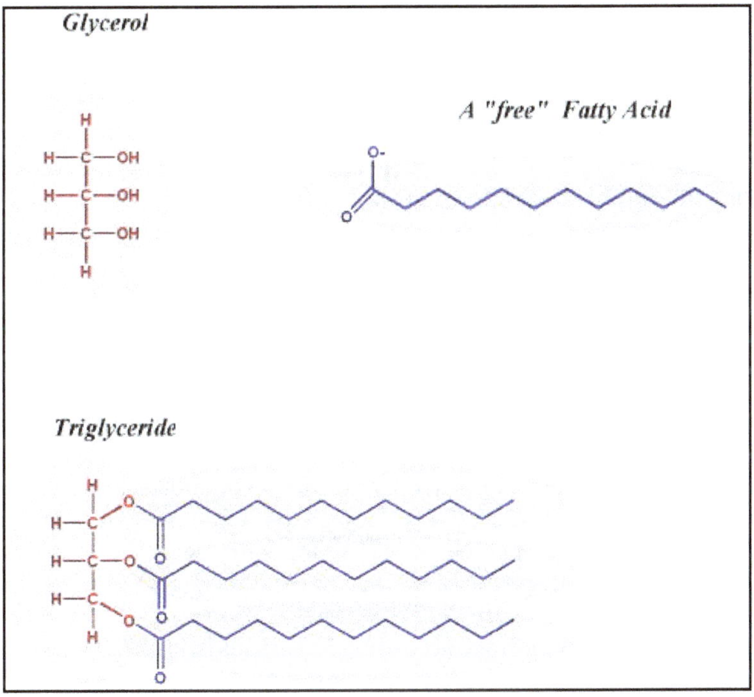

Glycerol

A "free" Fatty Acid

Triglyceride

European Official Methods of Analysis (EEC, 1991) is recommended to use for the determination of free fatty acid value in terms of per cent oleic acid. Approximately 1 g of potassium hydrogen phthalate ($KHC_8H_4O_4$) is weigh and dried in an oven at 110° C for 2 hours. Accurately 0.4 g of potassium hydrogen phthalate is weighed into an Erlenmeyer flask. About 75 mL of deionized water and 3 drops of phenolphthalein indicator (0.5 g phenolphthalein in 50 mL 95 per cent ethanol (v/v)) added into the flask before titration with potassium hydroxide

(KOH). Prepare 1 mol/L potassium hydroxide (KOH) with deionized water and standardize with potassium hydrogen phthalate. Prepared 50 mL mixture solution of 95 per cent ethanol-water solution (95:5 v/v) and 50 mL of diethyl ether (1:1 v/v). Add 3 drops of phenolphthalein indicator into the mixture. Titrate it with the ether-ethanol mixture with KOH solution until a sudden change in colour. Weigh 20 g of olive oil sample. The titrated ether-ethanol mixture is added to the 20 g of sample and 3 drops of phenolphthalein indicator added into the mixture before titration. Then, the mixture is titrated with mol/L solution of KOH and the volume of solution spent is recorded. Acidity is expressed as percentage of oleic acid with the equation given below:

$$V \times c \times \frac{M}{1000} \times \frac{100}{m} = \frac{V \times c \times M}{10 \times m}$$

where,

V = the volume of titrated KOH (mL)

c = exact concentration of the titrated solution of KOH (mol/L)

M = the molar weight of the oleic acid (282 g/mole)

M = weight of the sample (g)

4. Determination of the Acidity

The acidity expresses the percentage content (in weight) of the free fatty acids in the olive oil sample. Free fatty acids are normally present also in oils when the triglycerides are formed; there is a progressive increase in acidity due to the action of enzymes (lipase) naturally present in the olive fruit, which help the fatty acids to detach from the molecule of triglyceride (lipolysis). The lipolytic action of lipase produces free fatty acids which are responsible for the acidity of the oil. The same lipolytic action can be caused by enzymes produced by micro-organisms which grow on the fruit. Thus, in order to obtain a product which is organoleptically better and has lower acidity, it is necessary to preserve the olives well. Consequently, free fatty acid reflects the stability of oil and its susceptibility to rancidity. Acidity determination is mainly accomplished by titration using potassium hydroxide. The method determines the amount of free fatty acids (FFA) present in the oil, which is expressed as percentage of oleic acid. The free fatty acidity is a measure of the quality of the oil, and reflects the care taken in producing and storage processes of the oil. As well, acidity values are used as a basic criterion for classifying the different categories of olive oil. In contrast, according to Kiritsakis, *et al.* (2001), acidity is not considered as the best criterion for evaluating olive oil quality, since one oil with relatively high acidity may have a good aroma while another one with low acidity may not have so good a taste and aroma. For extra virgin olive oil the maximum acidity is 0.8 per cent, according to EU (European Union Commission, 1991).

Procedure

An appropriate sample amount is weighed into a 150 mL beaker, and then 50 to 100 mL solvent mixture is added. After stirring for 30 s the solution is titrated until the first equivalence point with alcoholic c(KOH) = 0.1 mol/L.

5. Peroxide Value

Peroxide value (PV) is a measure of total peroxides in olive oil expressed as miliequivalent of O_2 kg^{-1} oil (meq O_2/kg oil) and so this value is known as a major guide of quality. The official EU method is based on the titration of iodine liberated from potassium iodide by peroxides present in the oil. In other words, the peroxide value is a measure of the active oxygen bound by the oil which reflects the hydroxyperoxide value, and is one of the simplest measures of the degree of lipid peroxidation. The higher the number means the greater degradation due to oxidation. Peroxide value usually increases gradually over time after pressing. The upper standard value of the peroxide is 20 meq O_2/kg oil. In general, peroxide levels higher than 10 may mean less stable oil with a shorter shelf life (Nouros *et al.*, 1999).

Procedure

Weigh the test portion into a 250 mL Erlenmeyer flask with glass stopper. Add 50 mL of the 3:2 acetic acid-isooctane solutions. Swirl to dissolve the test portion. Add 0.5 mL of saturated K1 solution using a suitable volumetric pipette. Allow the solution to stand for exactly 1 min, thoroughly shaking the solution at least three times during the 1 min. Immediately add 30 mL of DI water. Titrate with 0.1 M sodium thiosulfate, adding it gradually and with constant and vigorous agitation. Continue the titration until the yellow iodine colour has almost disappeared. Add 0.5 mL of 10 per cent SDS. Add about 0.5 mL of starch indicator solution. Continue the titration with constant agitation, especially near the end point, to liberate all the iodine from the solvent layer. Add the thiosulfate solution drop wise until the blue colour just disappears. Conduct a blank determination of the reagents. The blank titrations must not exceed 0.1 mL of the 0.1 M sodium thiosulfate solution.

6. Specific Absorption Coefficients

They are also known as the UV Absorbance Values (K232 and K270). The specific absorption coefficients (specific extinction) in the ultraviolet region are needed for estimating the oxidation stage of olive oil. The absorption at specified wave lengths at 232 and 270 nm in the ultra violet region is related to the formation of conjugated diene and triene in the olive oil system, due to oxidation or refining processes. Compounds of oxidation of the conjugated dienes contribute to K232 while compounds of secondary oxidation (aldehydes, ketones *etc.*) contribute to K270 (Kiritsakis *et al*, 2000)

7. Determination of Anisidine Value

Anisidine value (AV) determination is an empirical test for assessing advanced oxidative rancidity of oils and fats. It estimates the secondary oxidation products of unsaturated fatty acids, principally conjugated dienals and 2-alkenals. Aldehydes are largely considered responsible for the off-flavours in fats and oils due to their low sensory threshold values. The AV test is particularly useful for oils of low peroxide value (PV) and for assessing the quality of highly unsaturated oils. The test involves a condensation reaction between the conjugated dienals or 2-alkenals and p-anisidine to form coloured products. The AV is empirically defined as 100

Table 23: European Regulation Standard Limit Values for Olive Oil Quality Parameters

Quality Indexes	*Acidity (Oleic acid per cent)*	*Peroxide Index (meq/kg)*	*K232*	*K270*
EVOO	≤ 1.0	≤ 20	≤ 2.5	≤ 0.20
VOO	≤ 2.0	≤ 20	≤ 2.6	≤ 0.25
Ordinary VOO	≤ 3.3	≤ 20	≤ 2.6	≤ 0.25
Lampante olive oil	> 3.3	> 20	≤ 3.7	> 0.25
Refined olive oil	≤ 0.5	≤ 5	≤ 3.40	≤ 1.20
Olive oil	≤ 1.5	≤ 15	≤ 3.30	≤ 1.00
Crude olive-pomace oil	> 0.5	—	—	—
Refined olive-pomace oil	≤ 0.5	≤ 5	≤ 5.5	≤ 2.5
Olive-pomace oil	≤ 1.5	≤ 15	≤ 5.3	≤ 2.0

times the absorbance of a solution resulting from 1 g of fat or oil mixed with 100 mL of isooctane/acetic acid/pansidine reagent, measured at 350 nm in a 10 mm cell under the conditions of the test. As the absorbance maximum shifts towards longer wavelengths with increasing unsaturation and as the colour intensity is greater with conjugated dienals than 2- alkenals, the absorption maximum varies from oil to oil. The AV obtained is only comparable within each type of oil. In spite of the low specificity of the AV test and has no a standard limit in olive oil codex, it is reported as a very useful indicator of oil quality and complements the PV test (Labrinea, *et al.* 2001).

It can be directly measured through flow injection FI analyzer consists of peristaltic pump, a spectrophotometer equipped with a Helma 181 flow cell (1 cm path length), a Rheodyne 5011P injection valve and an Advantech PCL-818 interface card for data acquisition and control. Glacial acetic acid, 2-propanol, isooctane, and p-Anisidine are used for estimation. A 40-1 aliquot of 10 per cent (w/v) olive oil samples in 2-propanol are injected in the p-anisidine in 2-propanol/acetic acid stream. Samples are mixed with the carrier reagent while flowing to the detector. The absorbance at 350 nm is continuously monitored resulting in typical FI, positive absorbance peaks.

8. Determination of Iodine Value

The Iodine number (IV) is defined as the number of grams of iodine which will add to 100 g of fat or oil. Iodine value shows the degree of unsaturation of the constituent fatty acids in an oil or fat and is thus a relative measure of the unsaturated bonds present in the oil or fat. Iodine value is expressed in grams of iodine absorbed by 100 g of oil or fat. Unsaturated compounds absorb iodine (in suitable form) and form saturated compounds. The amount of iodine absorbed in percentage is the measure of unsaturation in the oil. No oil has zero iodine value and oils are classified as drying, semi-drying and non-drying on the basis of iodine value. Oleic acid containing 1 double bond absorbs 90 per cent of iodine, linoleic acid (2 double bonds) absorbs 181 per cent iodine and linolenic acid (3 double bonds) absorbs 274 per cent iodine. Non-drying oils have 1 double bond and absorb iodine below 90 per cent. Semi-drying oils contain some proportion of double bonds and have iodine value below 140. Iodine value for coconut oil is 8, for olive 88, for human fat 105, for linseed about 200.

Chemicals Required

Wij's Solution: Dissolve separately 7.5 gm of AR Iodine tetrachloride and 8.5 gm of resublimed iodine in glacial acetic acid by warming on a water bath. Mix the two solutions and dilute to 1 litre with glacial acetic acid in cold.

Potassium iodide solution (15 per cent): Dissolve 15 gm of AR potassium iodide in 100 ml of water.

Sodium thiosulphate solution (0.1 N): Dissolve 25 gm of AR sodium thiosulphate crystals ($Na_2S_2O_3$ $5H_2O$) in a 1 litre of distilled water.

Starch indicator solution: 1 ml of starch in 100 ml boiling water.

Standardization of Sodium Thiosulphate Solution

Pipette out 20 ml of 0.1N potassium dichromate solution into a clean conical flask. Add 1 test tube of dilute H_2SO_4 and 10 ml of 15 per cent KI solution to the conical flask. Titrate against thiosulphate from the burette until it turned into pale yellow. Add 1 ml of starch indicator and titrate against thiosulphate solution. End point is disappearance of blue color.

Procedure

Weigh 0.5 gm of oil and transfer into Iodine flask. Add 10 ml of chloroform and warm slightly and cool for 10 minutes. Add 25 ml of Wij's solution in the same flask and shake vigorously. Then allow the flask to stand for half an hour in dark place. Add 10 ml of KI solution and titrate the solution against 0.1N Sodium thiosulphate solution until the appearance of yellow color. Add 1 ml of starch indicator and again titrate against the sodium thiosulphte solution. Disappearance of blue color indicates end point. Repeat the above procedure without taking sample (Oil) and note the corresponding reading for blank titration.

Formula for of sodium thiosulphate normality solution

$$N2 = \frac{V1 \times N2}{V2}$$

Formula for Iodine value calculations:

$$\text{Iodine value} = \frac{(V1 - N2) \times N1 \times \text{Equavalent weight of Iodine} \times 100}{\text{Weight of the sample}}$$

where,

V1 = Volume of thiosulphate required by blank in ml

V2 = Volume of thiosulphate required by sample in ml

N1 = Normality of thiosulphate

9. Colour Characteristics of Olive Oil

The colour is a sensory property with a strong influence on food acceptance as it contributes decisively to the initial perception that one can acquire of the condition, ripeness, degree of processing, and other characteristics of foods (Alos *et al.* 2006). As virgin olive oil is a natural product whose colour depends exclusively on biological compounds such as the chlorophyll and carotenoid pigments, their identification and individual evaluation make it possible to relate oil colour with the content and type of these compounds present. Oil appearance might be an indicator of a quality problem having occurred during blending, storage, crushing, and extraction or the refining process. The American Oil Chemists' Society (AOCS) has proposed four official methods for the colour determination of fats and oils. Related methods are Lovibond colour, Wesson colour, spectrophotometer colour and chlorophyll colour. Presently, CIELab, XYZ, Hunter Lab, and RGB (Red, Green, Blue) are the

Table 24: Major Pigments Found in Olive Oils

Olive Oil Samples	*Major Pigments found*
Verdial variety	Pheophytin a, chlorophyll a, chlorophyll b, pheophytin b, Lutein, β-carotene
Picual, Picudo, Subbetica, Hojiblanca, and Pajarero varieties	Pheophytin a, chlorophyll a, chlorophyll b, pheophytin b, Lutein, β-carotene, violaxanthin, luteoxanthin, antheraxanthin, mutatoxanthin, neoxanthin
Nevadillo, Hojiblanca, Picual, Martena, Pajarero, Arbequina, and Cornicabra varieties	Pheophytin a, chlorophyll a, chlorophyll b, pheophytin b, pheophorbide a, Lutein, β-carotene, violaxanthin, luteoxanthin, antheraxanthin, mutatoxanthin, neoxanthin, β-cryptoxanthin
Greek olive oils	Pheophytin a, chlorophyll a, chlorophyll b, pheophytin b, Lutein, β-carotene
Arbequina, Blanqueta, Cornicabra, Hojiblanca, Picual, and Lechýn varieties	Pheophytin a, chlorophyll a, chlorophyll b, pheophytin b, Lutein, β-carotene, violaxanthin, luteoxanthin, antheraxanthin, mutatoxanthin, neoxanthin, β-cryptoxanthin
Amfissis, Athinolia, Chondrolia, Kolovi, Koroneiki, Lianolia, Manaki, and Throumbolia cultivars	Pheophytin a Lutein, β-carotene
Miscellaneous	Pheophytin a, chlorophyll a, chlorophyll b, pheophytin b, Lutein, violaxanthin, neoxanthin
Verdial, Picual, and Manzanilla varieties	Pheophytin a, chlorophyll a, chlorophyll b, pheophytin b, Lutein, β-carotene
Arbequina variety	Pheophytin a, chlorophyll a, chlorophyll b, pheophytin b, pheophorbide a, Lutein, β-carotene, violaxanthin, luteoxanthin, antheraxanthin, mutatoxanthin, neoxanthin, α-carotene
Arbequina and Farga cultivars	Chlorophyll a, Chlorophyll b, Lutein, β-carotene, violaxanthin, antheraxanthin,neoxanthin
Cerasuola, Nocellara, and Biancolilla varieties	Pheophytin a, chlorophyll a, chlorophyll b, pheophytin b, pyropheophytine a, Lutein, β-carotene, violaxanthin, luteoxanthin,antheraxanthin, neoxanthin, β-cryptoxanthin
Arbequina	Pheophytin a, chlorophyll a, chlorophyll b, pheophytin b, pheophorbide a, Lutein, β-carotene, violaxanthin, luteoxanthin, antheraxanthin, mutatoxanthin, neoxanthin, α-carotene

alternative colour models that might be used in objective oil colour evaluation. L*a*b* is an international standard for color measurements, adopted by the Commission Internationale d'Eclairage (CIE). L* is the lightness component, ranging from 0 to 100, a* refers to the colour ranges from green to red and b* displays the colours from blue to yellow. These two chromatic components range from -120 to 120.

10. Analysis of Sensory Properties of Olive Oil

In the international trade specific vocabulary is used for virgin olive oil for the purposes of the method. The olive oil may have negative as well as positive attributes.

Negative Attributes

- *Fusty:* Characteristic flavour of oil obtained from olives stored in piles which have undergone an advanced stage of anaerobic fermentation.
- *Musty:* Characteristic flavour of oils obtained from fruit in which large numbers of fungi and yeast have developed as a result of the oil being stored in humid conditions for several days.
- *Muddy:* Characteristic flavour of oil that has been left in contact with the sediment that settles in underground tanks and vats.
- *Winey:* Characteristic flavour of certain oils reminiscent of wine or vinegar. This Vinegary flavour is mainly due to a process of fermentation in the olives leading to the formation of acetic acid, ethyl acetate and ethanol.
- *Metallic:* Flavour that is reminiscent of metals. It is characteristic of oil which has been in prolonged contact with metallic surfaces during crushing, mixing, pressing or storage.
- *Rancid:* Flavour of oils which have undergone a process of oxidation.

Other Negative Attributes

- *Heated:* or Characteristic flavour of oils caused by excessive and/or prolonged.
- *Burnt:* heating during processing, particularly when the paste is thermally mixed, if this is done under unsuitable thermal conditions.
- *Hay-Wood:* Characteristic flavour of certain oils produced from olives that have dried out.
- *Rough thick:* pasty mouth feel sensation produced by certain oils.
- *Greasy:* Flavour of oil reminiscent of that of diesel oil, grease or mineral oil.
- *Vegetable:* Flavour acquired by the oil as a result of prolonged contact with water vegetable water.
- *Esparto:* characteristic flavour of oil obtained from olives pressed in new esparto mats.
- The flavour may differ depending on whether the mats are made of green esparto or dried esparto.

- *Earthy:* flavour of oil obtained from olives which have been collected with earth or mud on them and not washed.
- *Grubby:* flavour of oil obtained from olives which have been heavily attacked by the grubs of the olive fly (*Bactrocera oleae*).
- *Cucumber:* flavour produced when oil is hermetically packed for too long, particularly in tin containers, and which is attributed to the formation of 2-6 nonadienal.

Positive Attributes

- *Fruity:* Set of the olfactory sensations characteristic of the oil which depends on the variety and comes from sound, fresh olives, either ripe or unripe. It is perceived directly or through the back of the nose.
- *Bitter:* Characteristic taste of oil obtained from green olives or olives turning colour. Pungent Biting tactile sensation characteristic of oils produced at the start of the crop year, primarily from olives that are still unripe.

Refernces

Alos, E., Cercos, M., Rodrigo, M.J., Zacarias, L., Talon, M. 2006. Regulation of color break in citrus fruits. Changes in pigment profiling and gene expression induced by gibberellins and nitrate, two ripening retardants. *Journal of Agricultural and Food Chemistry* **54**: 4888- 4895.

EEC 1991. Commission Regulation (EEC) No. 2568/91 of 11 July 1991 on the characteristics of olive oil and olive-residue oil and on the relevant methods of analysis. *Official Journal* **248**: 1-83.

Kiritsakis, K., Christie, W.W. 2000. Analysis of edible oils. In: L. John, and R. Aparicio (Eds.), Handbook of olive oil. Gaithersburg (USA): Aspen Publishers, pp. 129-151.

Labrinea, E.P., Thomaidis, N.S. and Georgiou, C.A. 2001. Direct olive oil anisidine value determination by flow injection. *Analytica Chimica Acta* **448**: 201–206.

Nouros, P.G., Georgiou, C.A., Polissiou, M.G. 1999. Direct parallel flow injection multichannel spectrophotometric determination of olive oil peroxide value. *Analytica Chimica Acta* **389**: 239-245.

25

Disease, Insect and Pest

The pest of the olive can be logically classified in three groups: (1) insects, mites, nematodes, (2) diseases, and (3) rodents. The kind of pest and its life cycle are important factors in developing effective control measures. The salient features are described below:

Major Diseases

1. *Armillaria* Root Rot

Symptoms

This disease is caused by oak root fungus (*Armillaria mellea*). The disease infected olive trees have thin canopies and appear weak. This symptom often develops first on one side of the tree and then takes several years to spread on the whole tree. The bark and outer wood of the upper roots and crown show discoloration. Infected roots have white to yellowish fan-shaped mycelial mats between the bark and the wood. Dark brown to black rhizomorphs can sometimes be seen on the root surface.

Management

Disease is more prone to the area where oak trees previously grew. Avoid planting olive orchards where forest or oak woodlands have recently grown or where there is a history of *Armillaria* root rot. If trees are infected, the growth of the fungus may be slowed by drying out the crown and upper root area of the tree. No olive rootstocks are resistant and infected trees cannot be cured.

Cultural Control

- ☆ Remove soil from around the base of the tree to a depth of 9-12 inches.
- ☆ Leave the trunk exposed and keep the upper roots and crown area as dry as possible.

- ☆ During winter, provide drainage if necessary so that rain doesn't collect in the hole.
- ☆ Re-check the hole every few years to make sure it has not filled in with leaves, soil, and other matter; the hole must be kept open and the crown and upper roots exposed.

2. *Cercospora* Leaf Mould

Symptoms

It is caused by fungus *C. cladosporioides or Pseusocercospora cladosporioides*. This disease often occurs together with peacock spot. The first signs are grey blotches on the underside of the leaves, the top of the leaves will yellow and these leaves may fall, causing some defoliation in some cases. Fruit can also develop small, brown lesion spots and not mature uniformly.

Management

- ☆ Prune to open the canopy for improved airflow.
- ☆ Reduce nitrogen use to prevent excessive canopy growth.
- ☆ Avoid excessive irrigation.
- ☆ Copper can be applied before the start of spring or autumn rains.
- ☆ Ensure good coverage of leaves.

3. Peacock Spot

Symptoms

It is caused by fungus *Spilocaea oleagina*. It is also known as 'olive leaf spot' and 'Bird's-eye spot', 'Peacock spot', develops with high humidity and rain. This disease occurs only sporadically, particularly when wet weather occurs in spring. Peacock spot appears on leaves as sooty blotches that develop into black, circular spots about 2.5–12 mm in diameter. There may be a yellow halo around the spot. The pathogen also infects fruit and fruit stems, but lesions are observed most often on upper leaf surfaces of leaves low in the tree canopy. Leaves fall prematurely. When significant defoliation occurs, strong bloom fails to develop and crop production is substantially reduced. Twig death may occur as a result of defoliation, and productivity is eventually further reduced. The typical symptoms are the appearance of sooty blotches on the leaves that later become muddy green to black, often with a yellow halo. Often the leaves drop prematurely.

Management

- ☆ Prune to open the canopy for improved airflow.
- ☆ Reduce nitrogen use to prevent excessive canopy growth.
- ☆ Avoid excessive irrigation.

4. *Verticillium* Wilt

Symptoms

It is caused by the fungus *Verticillium dahlia.* This is a soil borne fungus, which affects the roots and attacks the vascular tissue of the tree. Initially one or more branches turn yellow and wilted, usually early in the growing season, however the tree eventually die. Darkening of xylem tissue, a key symptom for distinguishing *Verticillium* wilt in many crops is frequently not apparent in olives.

Management

It should be based on avoidance of the fungus and reducing inoculum level in the soil.

- ✰ Use disease free planting material.
- ✰ Avoid planting into ground previously planted to alternative hosts of the fungus. *i.e.* stone fruit, Brassica's, potatoes and tomatoes.
- ✰ Avoid inter row cropping with susceptible plants, *i.e.* clover.
- ✰ Avoid soil movement from infected areas to non-infected areas.
- ✰ Reduce inoculums levels before replanting by keeping the soil weed free and growing resistant plants *i.e.* grasses for several years.

5. *Phytophthora* Root Rot (*Phytophthora* spp.)

Symptoms

It is caused by seven different species of *Phytophthora,* usually where excessively wet soils, clay-panning or poor drainage was a problem. Causes root rots, stem and crown cankers. Leaves wilt, yellow and may drop. Trees may die suddenly or slowly decline over several years.

Management

- ✰ Avoid water logging and excessive irrigation in the orchard.
- ✰ Avoid soil movement from infected areas to non-infected areas.
- ✰ Treat with Ridomil Gold (Metalaxyl-M).
- ✰ Fungicides do not eliminate *Phytophthora* from the soil and treatment must be required at regular interval.

6. Charcoal Root Rot (*Macrophomina phaseolina*)

Symptoms

Unlike *Phytophthora,* this fungus appears to like drier soil conditions, particularly where plants have been water-stressed during summer, causes a root rot. Affected roots have typical black speckles on their surface.

Management

- ✰ Avoid water stressing plants.

7. Bacterial Stem Cankers and Dieback

Symptoms

It is caused by bacteria, *Pseudomonas syringae, Xanthomonas campestris,* and *Ralstonia solanacearum.* These bacteria enter plants through pruning wounds or where frost/cold injury cause stems tissue to crack or peel. Symptoms vary from slow decline of trees and tree death, to localised cankers around wound sites.

Management

- ✰ Avoid wounding trees, as this acts as an entry point for bacteria.
- ✰ Copper can be used as a protectant, but is not able to eradicate established infections.

8. Anthracnose

Symptoms

It is caused by bacteria *Colletotrichum acutatum.* This disease needs wet conditions with high humidity. It affects fruit close to harvest. Causes soft circular rots on the fruit, usually on the shoulder, and at high humidity produces an orange slimy mass of spores on the fruit surface.

Management

- ✰ Prune to aerate the canopy and copper based chemicals can be applied as a protectant, and there is also a permit for Amistar, which should also be applied as a protectant.

Several other diseases are also infect olives, which causes minor damage to mature trees but often significance to young plants. These includes:

- ✰ **Rhizoctonia root rot (*Rhizoctonia* spp.):** It usually infects roots of young plants and causes browning and rotting. Above ground symptoms include tip death, defoliation or death. While on roots of mature plants, it do not cause severe problem.
- ✰ **Stem cankers (*Botryosphaeria* sp.):** This fungus is occasionally detected on branches of trees, resulting in yellowing of foliage above the affected area. In Western Australia, the same fungus has been detected on apple and stone fruit trees, which may be responsible for cross infection of nearby olives. This fungus can also be a secondary coloniser of dead wood.
- ✰ **Minor root rots (*Pythium* spp. and *Fusarium* spp.):** These fungi are common in all soils, but are more prevalent in wet, poorly drained areas. They are not considered to be a major problem with mature trees, but will seriously affect young trees and those weakened by other stresses.
- ✰ **Fruit Rots (*Botryosphaeria* sp., *Alternaria* spp. and *Coleophomaoleae*):** Usually occur on fruit already damaged by other causes, particularly in wet and humid weather. Avoid damage to fruit.

- **Crown Gall (*Agrobacterium* sp.):** So far this has only been detected in potted nursery stock, but could be serious if it establishes in the field.

Insects

1. Olive Fruit Fly (*Bactroceraoleae*)

Olive fruit fly is posing a serious threat to the olive industries. This insect is native of Eastern Africa; it is considered the most damaging pest of olives all over the world. The adult olive fruit fly is about 4–5 mm long with clear wings containing dark veins and a small dark spot at the wing tip. The head, thorax, and abdomen are brown with darker markings, and the thorax has several white or yellow patches on each side. The end of the male fly's abdomen is blunt, whereas females have a large black ovipositor at the end of their abdomen that is visible to the naked eye. Larvae are yellowish white maggots with a pointed head. Mature larvae pupate in fruit in summer; in fall they leave the fruit and pupate in the soil under the tree. Damage by olive fruit fly includes oviposition "stings" on the fruit surface, fruit drop, or direct pulp destruction rendering fruits useless for canning. Larval feeding allows microorganisms to invade the fruit, causing rot and lower oil quality. In areas of the world where olive fruit fly is established and not controlled, its damage has been responsible for losses of up to 80 per cent of oil value because of lower quantity and quality, and in some varieties of table olives, this pest is capable of destroying 100 per cent of the crop.

Management

- Sanitation is important in reducing overall fly densities.
- Remove old fruit remaining on trees following harvest and destroy all fruit that are on the ground by either burying at least 4 inches deep or taking to the landfill.
- Prevent fruiting on landscape trees in spring by using a chemical like "Fruit Stop" or destroy fruit on the ground in fall to reduce this invasion pathway.
- Olive fruit fly adults feed on honeydew. Reducing black scale populations may reduce a food source needed during high summer temperatures.
- Cultural controls, the use of GF-120 Fruit Fly Bait, sprays of kaolin clay, and mass trapping are acceptable for use in an organically certified crop.
- *P. concolor,* is a parasite that can be raised in culture and has been released for other fruit flies including the Mediterranean fruit fly, have been attempted in California with limited success to date.

2. American Plum Borer (*Euzophera semifuneralis*)

Larvae of this insect attack soft, spongy, callus like tissue, which occurs at graft unions, tree wounds, and in olive knots. They can also feed into normal tissue, girdling limbs, which tends to break. Gum liquid exudate and frass can be seen around injured wood. The adult borer is gray in colour with brown and black markings on the wings however mature caterpillars are dusky white or pinkish and are about 1 inch long. Adult females lay eggs near where callous tissue has developed, such as at pruning wounds, crown galls, or scaffold crotches. Larvae bore into the tree to feed on vascular tissue. American plum borer overwinters in a protective cocoon spun in a sheltered location on the tree; pupation takes place in spring. There are three to four generations each year.

Management

- ☆ Monitor trees in spring and summer for gum pockets and frass.
- ☆ The borer can be detected by brownish frass and webbing at feeding sites.
- ☆ If larvae are present, remove and destroy infested wood if possible.
- ☆ If wood cannot be removed, spray trees with a hand held sprayer from one foot above the scaffold crotch to one foot below, two to three times during the growing season.
- ☆ The first application should be mid- to late- April and subsequent applications at 6-week intervals.
- ☆ Efficacy is improved if the trunk is painted immediately following a trunk spray with a latex paint to protect against sunburn.
- ☆ The paint helps to preserve the insecticide and give protection over a longer period of time.

3. Black Scale (*Saissetiaoleae*)

Black scale prefers dense and unpruned portions of trees. Open, airy trees rarely support populations of black scale. Young black scales excrete sticky, shiny honeydew on leaves of infested trees. At first, affected trees and leaves glisten and then become sooty and black in appearance as sooty mold fungus grows on the honeydew. Infestations reduce vigour and productivity of the tree. Its continued feeding reduces bloom the following year. Olive pickers are reluctant to pick olive fruits covered with honeydew and sooty mold. The adult females are about 5 mm in diameter and dark brown or black with a prominent H-shaped ridge on the back. Young scales are yellow to orange crawlers and are found on leaves and twigs of tree. A hand lens is usually needed to detect the crawlers. Black scales are in the soft scale family (*Coccidae*) and usually have one generation per year in interior valley olive-growing districts. In cooler, coastal regions multiple generations occur.

Management

- ☆ Pruning to provide open, airy trees discourages black scale infestation and is suitable to chemical treatment.
- ☆ A number of parasites attack black scale, the most common are *Metaphycushelvolus*, *M. bartletti*, and *Scutellistacaerulea* (=*S. cyanea*).
- ☆ These parasites, combined with proper pruning, provide sufficient control.
- ☆ Control ants using bait stations in the orchard because they disrupt biological control.

4. Branch and Twig Borer {(*Melalgus* (=*Polycaon*) *confertus*)}

Adults bore small, round holes at the base of buds or axils of twigs injured by sunburn. Eggs are laid at these locations in early May and grubs bore into the heartwood, where they live for a year or more. Twigs break at location of the injury. Madrone, oak and grape are the preferred host plants. The adult borer is a 7–15 mm long beetle, mostly black with brown wing covers. The C-shaped, white larvae are covered with fine hair. There is one generation per year.

Management

- ✰ Prune out and burn infested wood.
- ✰ Prevent sunburn and other injury that predisposes trees to damage.

5. Oleander Scale (*Aspidiot usnerii*)

The oleander scale infests olive fruit and delays maturity at the spot where it feeds. Thus, damage is seen as prominent green spots on purple fruit, in direct contrast to the dark spots caused by olive scale. Heavy infestations seriously deform fruit and fruit spotting renders the olive worthless. Extremely heavy infestations reduce oil content by as much as 25 per cent. Leaf and twig damage also result in loss of the production. The adult female is an armoured scale that is about 2.5 mm long and oval. It has a waxy covering that is whiter than olive scale with a yellow or light brown spot near the centre. The adult male scale is elongate. If the coverings are removed, the female body is yellow, while the male scale is brownish yellow. This scale is most common on leaves in the lower part of the tree. There are several generations a year.

Management

- ✰ Scale can be effectively controlled by natural enemies (several species of *Aphytis*, including *A. melinus*, are important parasites of oleander scale) and does not usually cause economic damage.
- ✰ Preserve these natural enemies by selecting insecticides for other pests that do not kill beneficial insects.

6. Olive Mite (*Oxyenus maxwelli*)

The olive is the preferred host plant for olive mite insect. However, varieties like 'Ascolano', 'Sevillano', 'Manzanillo', and 'Mission' show high to low susceptibility to this pest. The olive mites feed on succulent stem and bud tissues and on the upper surface of leaves. Gross symptoms of mite damage include sickle-shaped leaves, dead vegetative buds in spring, discoloration of flower buds, bud drop, blossom blasting, inflorescence abscission and reduced shoot growth. It is an eriophyid mite and is difficult to see without magnification. The mite is yellowish to dark tan, slow moving, and has a wedge-shaped body that is typical of many eriophyid species.

Management

- ✰ Olive mite is generally not managed in olives unless fruit set and crop have been below normal for several years.

- ☆ If crop yield has been increasingly poor for several years in a row, examine shoot tips and developing flower buds in spring for the presence of olive mites.
- ☆ Treat before bloom if large populations are present.

7. Olive Psyllid (*Euphyllura olivine*)

In severely affected orchards it can cause yield losses of 30 to 60 per cent. The insect directly feed on buds, flowers, tender shoots, and small fruit and also through the production of honeydew, which increases sooty mold development. During olive flowering and fruiting, psyllid waxy secretions cause flower and small fruit drop and yield reductions. Large populations may retard the growth of young trees. It feeds on olive (*Olea europea*), Russian olive (*Elaeagnus angustifolia*) and mock privet (*Phillyrea latifolia*). Light green and tan adults are 2.5 mm long and strong jumpers. They don't move very far, leading to clumped distributions. In warm temperatures (above 80°F), psyllids become inactive. Mortality may occur at temperatures above 90°F.

Management

- ☆ Pruning of limbs is good to enhance air circulation and reduction of insect attack.

8. Olive Scale (*Parlatoriaoleae*)

In the early growing season during late May to June, the olive scale feed on and consequently deforms young, rapidly growing fruit. A later brood, in July and August, causes the pronounced purple spotting of green fruit, rendering it worthless for most markets except perhaps black ripe process. Heavy olive scale infestations will also occur on branches, twigs, and leaves. Such infestations substantially reduce the productivity of a tree. The insect have armored scale and like all armored scales, resembles a small encrustation on the plant. The adult female scale is about 2.5 mm long, with a grayish, oval, waxy covering. The male scale is more elongate with a black spot at one end. If the coverings are removed, the scale bodies of both sexes are reddish purple. It feeds on twigs, leaves, and fruits. However, it is most often noticed at harvest; dark purple spots occur on otherwise green to yellowish fruit where the scale has settled. There are several generations a year.

Management

- ☆ Olive scale can be effectively controlled by natural parasites, *Aphytis maculicornis* and *Coccophagoides.*
- ☆ Chemical treatment is rarely needed for olive scale unless biological control is disrupted by treatments applied for other pests.

26

Major Physiological Disorders

Physiological or abiotic disorders are distinguished from other disorders in that they are not caused by living organisms (viruses, bacteria, fungi insects etc), but are caused by non-living, abiotic situations and cause a deviation from normal growth. They are physical or chemical changes in a plant from what is normal and generally caused by an external factor. Some non-infectious disorders are easy to identify, but others are difficult or even impossible. Most of them are not reversible once they have occurred. To help in identifying physiological disorders it important to know that:

1. Physiological disorders are often caused by the lack or excess of something that supports life or by the presence of something that interferes with life.
2. Physiological disorders can affect plants in all stages of their lives.
3. They occur with the absence of infectious agents therefore cannot be transmitted.
4. Plant reactions to the same agent vary widely, from little reaction to death.
5. Dealing with physiological disorders often means dealing with the consequences from a past event.
6. There is generally a clear line of demarcation from damaged and undamaged tissue.
7. Physiological disorders are serious in themselves but often serve as the 'open door' for pathogens to enter.

The given below physiological disorders have been observed in olive:

1. Nail Head

This disorder is characterized by surface and spotting. It results from the death and collapse of epidermal cells, which create air pockets underneath the fruit skin.

Symptoms are observed on fresh olives kept at 10°C for six weeks or longer or 7.5°C for 12 weeks or longer. Delay in placing the olives in brine after harvest may cause nail head. These depressions persist in the pickled product and are thought to be caused by bacterial action, as colonies of bacteria are found in them. Nail head is avoided by pickling olives promptly after harvest or storing them promptly in brine.

2. Carbon Dioxide Injury

Internal browning and increased decay incidence and severity results from exposure to more than 5 per cent CO_2 for longer than four weeks. CO_2 injury is evidenced by internal browning and increased decay incidence and severity.

3. Chilling Injury

The incidence and severity of chilling injury (CI) on fresh olives depend on storage temperature and duration as well as cultivar. The order of susceptibility to CI is Sevillano (most susceptible) - Ascolano - Manzanillo - Mission (least susceptible). It can be a major cause of deterioration if fresh olives are stored before processing for longer than 2 weeks at 0°C (32°F), 5 weeks at 2°C (36°F), or 6 weeks at 3°C (38°F). Internal browning begins in the flesh around the pit and radiates outward toward the skin as time progresses. Skin browning indicates an advanced stage and/or greater severity of CI. CI stimulates respiration and ethylene production rates of fresh olives. Exposure to CO_2 levels above 5 per cent aggravates CI, while 2 per cent O_2 is beneficial in maintaining flesh firmness and green color of the skin in olives kept at 5°C (41°F) or higher temperatures. To manage CI avoid exposure of fresh olives to temperatures below 5°C (41°F). Ideal storage conditions are 5 to 7.5°C (41 to 45°F) and 90-95 per cent relative humidity (Kader *et al*, 1990).

4. Apical End Rot

Also known as apical end desiccation, soft nose. The apical (blossom) end of the fruit shrivels, mostly close to maturity. The internal flesh and pip may be blackened, either at the apical end only or throughout the whole fruit. Sometimes secondary fungal rots infect the shrivelled end. The cause is unknown. It may result from sudden changes in temperature and humidity, which produce partial dehydration of the fruit at the apical end. It has also been associated with calcium and boron deficiencies, and with changes in watering regimes.

5. Clay-Panning and Root Plaiting

Clay-panning and root plaiting are disorders in root architecture that can lead to unthrifty plants that are subject to stress-related dieback and infections. Clay-panning is caused by poor soil structure and ground preparation, whereby a hard layer of subsurface soil prevents roots from growing downwards. Affected trees may also be subject to temporary water logging, which can lead to further disorders and infections. Conversely, dry soil can exacerbate stresses, because plants cannot draw moisture from deeper in the soil. Trees may also be subject to blowing over in strong winds because of their poorly anchored root systems. Root plaiting occurs when plants become pot-bound during their nursery production. Plants have reduced and weaker root systems, which allow environmental stresses to lead to various disorders and infections.

6. Sphaeroblasts

Sphaeroblasts are knob-like growth up to 10 mm wide which protrude from stems. When they are cut open, a spherical lump of wood can be removed from the surrounding tissue. Their cause is unknown, and they commonly occur on the cultivar 'Barnea'.

7. Oedema

Symptoms of oedema are small, brown, corky growth up to 5 mm wide that form on the surface of stems or roots from enlarged lenticels (breathing holes in bark). They occur when high soil moisture causes excessive water uptake, which engorges the cells near the lenticels. These cells can rupture from the high water pressure, and the plant forms callus tissue in an effort to heal. When roots experience periods of high soil moisture, some tissue may also asphyxiate (because of reduced oxygen levels). Consequently, these roots are predisposed to infection by a range of minor pathogens or opportunistic invaders such as *Fusarium*, *Pythium* and various bacteria.

8. Stem Death

The stem of the plant dies a few centimetres above ground level. The base is generally healthy and new shoots will appear below the dead stem. Most common on young trees in their first and second years in the field. The cause is Unknown mostly damage occurs to the young tree and allows entry of wound-invading bacteria and fungi. Damage is often associated with cold temperatures, sun scald and herbicide, but in many cases the cause has not been determined.

9. Tip death

Symptoms start with ends of branches which die for no apparent reason. Tip death appears to have no effect on the general health of the tree or its productivity. Branches can be removed if this is considered necessary for cosmetic purposes. Inspection of the stem below the dead tips is needed to determine whether the death has a specific c cause which should be further investigated. Root rot and trunk cankers can also cause tip death, and so should be investigated.

10. Miscellaneous Leaf Damage

There are many other symptoms seen on leaves that have no known cause. They may result from infection by biotic agents or from environmental or nutritional effects. Symptoms include white spotting, pale brown blotches, striping and yellowing, and dead tips.

References

Kader, A.A., Nanos, G. D. and Kerbel. E.L. 1990. *Storage potential of fresh 'Manzanillo' olives. Calif. Agr.* **44**(3): 23-24.

27
Impact of Abiotic Stresses in Olive

The abiotic stresses like lack/excess of essential nutrients or antagonism among them, excess of nonessential elements, unsuitable environment (too cold or hot, too wet or dry, or too windy) and soil characteristics such as poor physical conditions, improper pH, *etc.* Environmental pollution and climatic extremes like lightning, hail, and snow) and their interactions with plant health, production, and quality can greatly influence the production and productivity of olive.

Climatic Conditions

Climatic conditions are limiting factors in the distribution of olive throughout the world. Commercial olive-growing areas are not found above a latitude of 45° north and south as olive generally succumbs at -10 to -12°C (Mancuso, 2000). Temperature and other climatic factors have a range which maximizes growth, so that when they are outside this range, the different physiological processes are slowed down or suspended.

Low Temperatures

Low temperatures are determinant for the successful cultivation of olive, since it can influence the physiology of the plant by slowing down many vital processes, physiological requirement of chilling temperature to break dormancy of flower buds, chilling injury, or by inducing frost damages and physiological drought.

Chilling Injury

Chilling injuries can occur at temperatures between 15°C and 0°C, depending on the vegetative stage of the trees. At chilling temperatures, the plant slows down its metabolic processes, reducing respiration and enzymatic activity with a slower absorption of water and nutrients and lower photosynthetic efficiency and cellular processes; plants stop growing and the young leaves are pale-green. Chilling injury becomes evident in marginal areas with long periods at low temperatures.

Frost Damage

Frost damage occurs at temperatures lower than 0°C with the formation of ice. The threshold temperature below which olive shows the first metabolic effects of cold damage is not easily determined, since it depends on many factors, such as growth stage and age of the plant, its nutritional and sanitary status, the type of organ affected. The velocity at which minimum temperature is reached and its duration, the site exposition of the grove, the air and soil humidity and the presence of weeds and wind. Furthermore, the damage depends also on the time of the year when the frost is likely to occur, being classified as early frosts, winter frosts and late frosts, occurring in fall, winter and spring, respectively.

Early Frosts

A drop in temperature during fall is generally of limited importance since temperature reached is generally not harmful for leaves and twigs. Low temperatures, such as -0.4°C, can induce dehydration of the drupes and skin shrivelling, which may represent serious damage mainly for table olive (Azzi 1928) and a temporary reduced accumulation of oil. However, at temperature lower than -1.7°C fruit can be extensively damaged showing surface blister and spots. Damaged drupes turn brown, acquire an aqueous consistency and drop, or dehydrate remaining shrivelled until harvest.

Winter Frosts

In general, during winter olive plants are damaged by temperatures lower than -7 to -8°C, although there are cases of resistance to -12°C and, much less frequently, at -18°C. Depending on the low temperature reached and its extent, the damage can affect leaves, twigs, branches, and the trunk of the plant. The temperature at which lethal intracellular ice is formed in cold-stricken plants differs according to the organ in decreasing order of sensitivity: drupes > roots > leaves > twigs > buds (Graniti *et al.*, 2011). Symptoms of frost damage are as follows:

1. Appears first on the leaves which desiccate, turning brown due to the oxidation of phenols leaked in the intercellular spaces. Generally, after some days, the damaged leaves fall and the plant, in relation to the age and the severity of frost, may be completely defoliated.
2. Ice formation in the leaf tissues does not necessarily lead to immediate defoliation, which may occur after a period of transition (also 1-2 months), in which leaves show different hues of colour.
3. Cracked and detached bark can be observed mainly on the trunk and scaffold branches, even a few hours after the frost. The external part of the xylem, in contact with the bark, changes colour from cream to dark-brown, due to the release and oxidation of phenols, sometimes only the cambium turns dark brown, whereas the bark and the xylem appear apparently normal.

4. Cortical tissues detach at the cambium level, but in some cases only the transparent epidermis comes off, revealing the underlying green bark tissue (Gucci *et al.*, 2003).

Late Frosts

In the spring, during the new flush of growth, olive tree become very susceptible to frost injury and can suffer even at temperatures just below freezing. Symptoms of late frost damage are as follows:

1. Direct damage is observed first on more hydrated tissues, such as vegetative apexes and young leaves which desiccate, turning brown.
2. Leaves with light symptoms of late frosts may show a pale green colour, smooth areas or patches of the lower surface, that may cover most of the blade (Furthermore, whole flower clusters as well as single flowers and their pistils in particular, that seems very sensitive, may be injured.
3. Low temperatures can be one of the factors determining the production of pseudo drupes, also known as "shot berry" (Michailides *et al.*, 2011).
4. Pseudo-drupes are characteristically small, almost spherical fruits that develop in clusters, in which the pericarp is complete whereas the endocarp contains pseudo seeds showing degeneration of the four ovules and no embryo (Messeri, 1947).

High Temperatures

High temperatures are less harmful than low temperatures in both pathological and ecological contexts. High temperature damage is rare and occurs only in coincidence with drought, excessive light and strong winds accompanied by low humidity. Olive cultivars acclimated to high temperatures (*e.g.* 40°C) still maintain 70-80 per cent of their photosynthetic rate, thus their cultivation is expanding in rainfed and desert areas. Damage due to high temperatures is expected to rise, considering the general accumulation of greenhouse gases in the earth's atmosphere which are leading to an increase in the average temperatures (Gibelin and Déque 2003).

Snow

Snow is major problems in temperate region especially at higher latitudes and altitudes. Breakage or stretching of scaffold branches due to the snow weight can cause extensive damage, especially on plants and cultivars rich in leaves and twigs like those neglected in pruning. During the day a snow covering is disadvantageous since it keeps tree parts exposed to sunlight colder than when they are not covered. Sunlight cannot heat the peel because the reflectivity of fresh snow to visible or near infrared is high. Emissivity and thermal conductivity of the snow, when dry, are very low, and therefore during the night the snow acts as an insulator against radiative loss of heat to the cold, dry sky. For the same reason, snow can protect the canopy from very cold air drift and the soil from loss of heat. However, these advantages are lost when snow melts, for its thermal conductivity augments.

Moreover, evaporation of melted snow increases the loss of heat during the night, leading to frost damage on the sky-exposed side of the branches.

Drought

Olives are generally drought tolerant. This is due to morphological, physiological, and biochemical adaptation of the trees that reduce water loss and maintain water uptake at a high plant water status as the drought commences; moreover, they maintain turgor and tolerate dehydration at low plant water status as the drought persists. Olive plants combine low leaf conductance, leaf area and radiation load, deep roots, high root length density, and high hydraulic conductance; turgor is maintained by osmotic adjustment of cell contents, small cell size, changes in cell wall elasticity and dehydration is tolerated also because of other properties of the protoplasm and cell wall. However, in spite of the well-recognized tolerance to water shortage, olive groves may suffer when persistent drought and high temperatures occur simultaneously.

Drought Consequences

1. Drought stress has a negative effect on the growth parameters of olive trees, particularly in young plantations.
2. Shoot growth is very sensitive to water shortage and is economically significant, since it has an effect on the non-bearing fruit period of the trees.
3. Trunk expansion is also reduced. The reduced growth rates is attributed to the lower photosynthetic rates due to stomatal closure and also to the reduced functionality of the root system, which is unable to supply the canopy with minerals and water.
4. Drought also decreases the actual water content and succulence of the leaves, carbon assimilation rate, stomatal conductance and intrinsic water use efficiency, while leaf tissue density and intercellular CO_2 increase.

Wind

Olive trees often show symptoms of wind damage because they are frequently cultivated in hot arid climates where the effect of the wind *per se* is combined with low soil water availability and relative humidity. Moreover, sometimes olive groves are located on unfavourable hill sites, constantly hit by wind, or in areas bordering the sea thus exposed to salty winds. Wind damage varies with its intensity and/ or duration, temperature, air and soil humidity, presence of polluting substances (*e.g.*, salts). Autumn winds, especially in combination with drought, may lead to shrivelling of the drupes, which is particularly damaging for table olives. Windstorms may induce fruit bruising and dropping, breakage or stretching of the branches, and also uprooting in the case of plants grown from cuttings. Constant winds, especially coming from the sea and thus also rich in salts, may lead to a limited growth and canopy asymmetry (flag canopy).

Hail

Depending on when a hailstorm occurs, the intensity of the event, and the hailstone size, hail can damage foliage, flowers, fruits, tender stem tissues and the epidermis of twigs, branches, and trunk. Symptoms can be seen over a broader area and consist of severe defoliation, leaf, flower and stem lesions, scars and bruises on fruit and branches. Lesions on the fruits can make table olives unmarketable because of the presence of blemishes and malformations.

Lightning Injury

Trees shows often have a strip of bark and sapwood blown off the trunk, leaving a continuous or intermittent rough groove which follows the grain of the wood. Trees that are struck but not killed are likely to be disfigured by the death of limbs. In addition, wounds and destroyed parts provide entry for wood decay fungi. While in traditional training system injury is observed on a single plant, in the new ones (high and super high density growing system) damaged plants can be observed in a row, thanks to the wires connecting them.

Water-logging

Water logging alters metabolic functions of the roots resulting in reduced absorption and transport of water and nutrients.

Symptoms Under Water Logging

1. Stressed plants shows a reduced growth and a chlorotic condition of the leaves, which then turn brown and are shed.
2. Young stressed plants may show curled and twisted dry leaves. Flowering, fruit set and growth of the drupes are negatively influenced by the stress.
3. Fruit of flooded plants show brown spots or a complete discoloration, involving also the still unlignified endocarp.
4. The root system appears damaged, showing a few and discoloured rootlets; in severe cases, the soil emanates an unpleasant smell.
5. Floods may contribute to the dissemination of soil-borne pathogens from infested to healthy areas and may predispose roots to soil-borne pathogen attack.

Salinity

Salinity, either occurring naturally or induced by irrigation, is of concern in many semi-arid areas to which olive is climatically adapted. The salt most concerned is NaCl, but also SO_4^{--}, HCO_3^-, and CO_3^{--} ions may contribute to increase salinity damage. In olive plants, salinity induces a strong reduction of leaf fresh weight and oxidative stress, and a dramatic increase in proteins that undergo tyrosine nitration (Valderrama *et al.*, 2007); growth is negatively affected by restricted availability of water, toxic effects of Na^+ and Cl^-, and the metabolic energy expended in salt exclusion by roots and/or its sequestration within the plant (Connor and Fereres, 2005).

Nutrient Deficiency and Excess

Nitrogen (N)

N is a constituent of amino acids, amides, enzymes, alkaloids, chlorophyll, and other compounds important in plant metabolism, especially proteins, it directly influences photosynthetic capacity, carbon assimilation, dry matter accumulation, fruit set, shoot growth, and yield. Symptoms of deficiency may appear on canopy when leaf N content is lower than 1.4 per cent dry weight, whereas 1.5-2.0 per cent N is considered adequate (Fernández-Escobar, 1998; Reuter *et al.*, 1997).

N Deficiency Symptoms

1. The life of a leaf may be shortened, and on the whole the canopy may look thin; such trees suffer from exposure to sun and are prone to sunburn.
2. Twigs may die back and fruit production may be limited in quality and quantity, being N deficiency associated also with pistil abortion.

Phosphorus (P)

Phosphorus has important functions in plants since it is a constituent of nucleic acids and of membrane phospholipids, and also participates in energy transfer and in regulating carbohydrate metabolism. This element is readily translocated and concentrated in metabolically active tissues Due to its mobility in the plant, the relatively low levels needed, and the intrinsic characteristics of olive plants.

P Deficiency Symptoms

1. These are rarely found anywhere in olive the world. However, when present, symptoms are very similar to those of N deficiency, with plants showing reduced growth and yield.
2. Leaves show a reduced size but are not deformed or chlorotic as in N deficiency.

Potassium (K)

Potassium is a highly mobile nutrient both within the cell and through the xylem and phloem, thus deficiency are appears on older leaves. A content less than 0.4 per cent (dry weight) can induce symptoms of deficiency, whereas a content of over 0.8 per cent is considered adequate. (1) Play important role in buffering anions and in the stabilization of cell pH, activator of many enzymes in meristematic tissues and participates in protein synthesis and photosynthesis. (2) Plays an important role in water balance in the plant since it is required for turgor build up and for maintaining the osmotic potential of cells, which in guard cells governs the opening of stomata. (3) It is involved in carbohydrate metabolism, water uptake from the soil, water retention in the tissue and long distance transport of water and metabolites in the phloem and xylem.

Deficiency Symptoms of K

1. First appears on the leaves consist of a slightly reduced size and a chlorosis of tips and edges, that develop into necrosis looking like "burns". These symptoms may resemble those of boron deficiency.
2. The areas of dead tissue progress from the tip to the base of the leaves and from the margin towards the interveinal space.
3. The leaf tip tends to curve downwards; twigs show short internodes.

Magnesium and Calcium (Mg, Ca)

Calcium and magnesium are closely related elements and their chemistry is somewhat similar. Magnesium is an essential part in the chlorophyll molecule and is a co-factor in a number of enzymes including transphosphorylase, dehydrogenase, and carboxylase; it aids in sugar and lipids biosynthesis and activates the formation of the polypeptide chain from amino acids.

Deficiency Symptoms

1. First appear primarily on mature leaves, containing less than 0.1 per cent (dry weight) Mg, and show as pale green colouring on the half apical part of the leaves, which gradually extends the whole blade.
2. In severe cases, leaf shedding and poor vegetation may complete the picture.
3. Terminal leaves are normal in size and colour; fruit can acquire a chlorotic appearance but less pronounced than in iron-deficient trees.
4. The main symptom of Ca deficiency is a chlorosis of the apical part of the leaves with the veins in the chlorotic area of older leaves becoming almost white.
5. Root system may show poor growth, and root apexes may appear gelatinous.

Zinc (Zn)

Zinc acts either as a metal component of enzymes or as a functional, structural, or regulatory cofactor of many enzymes. It is transported in the xylem as free cation or bound to organic acids and it is a component of auxin molecules, one of the best-known enzymes regulating plant growth In some crops Zn deficiency often accompanies other metal deficiencies (*e.g.* Zn and Mn or Zn and Fe), but in olive these combinations have never been reported. Under experimentally-induced conditions of Zn deficiency plants and shoots appeared normal, whereas younger leaves showed a slightly lighter colour compared to older leaves and the fruits matured earlier.

Iron (Fe)

Iron is an essential element for the synthesis of chlorophyll and is involved in nitrogen fixation, photosynthesis, and electron transfer. It is also a structural component of substances involved in oxidation-reduction reactions, reduction of O_2 to H_2O during respiration in particular, and of respiratory enzymes as a part of cytochromes and emoglobin, and in many other enzyme systems. Iron deficiency often results from a high pH or mineral imbalance which renders Fe unavailable, rather than from an actual lack of iron in the soil.

B Deficiency Symptoms

1. Affected trees show an overall loss of vigor, general yellowing or chlorosis of the leaves and low productivity.
2. The veins, which remain green and stand out in contrast with the pale yellow blade, are the most remarkable and distinctive features of this deficiency.

Boron (B)

Boron is an essential element absorbed by roots as the very weak acid H_3BO_3 and is transported through the xylem. It affects carbohydrate metabolism and transports across cell membranes, playing a role in amino acid formation and synthesis of proteins and cell wall material Because of its impact on cell development and on sugar and starch formation and translocation,

B Deficiency Symptoms

1. It retards the new growth and tree development.
2. It apparently leads to a build-up of indoleacetic acid (IAA) by blocking IAA oxidase in the roots, which inhibits elongation; it also reduces membrane stability, pollen germination and the growth of pollen tubes (Bennet, 1993).
3. Boron is not very mobile, which explains the appearance of deficiency first on young leaves, the tip of which turns light green (chlorotic tips), then during fall and winter most of the leaves turn yellow with shades of bronze and eventually desiccate.
4. Trees suffering from B deficiency appear chlorotic from a distance, exhibit a bunchy growth and delayed bud break (Ciccarone, 1956).
5. In severe cases, a dieback of most of the upper twigs occurs, which is followed by the growth from lower nodes of many secondary shoots (excess branching) with short internodes, that lead to the formation of rosettes.
6. The bark of affected twigs, branches, and trunks may show protuberances 1-2 mm high and 5-10 mm long either smooth or, sometimes, with a cracked surface.
7. The phloem underneath these protuberances shows variously extensive necroses.
8. Depressed, corky and smooth reddish lesions can also be observed on branches and trunk.

Pollutants

Over the past few decades there has been a steady and significant increase in water, soil, and air pollution caused by the release of pollutants produced by human, industrial, and agricultural activities. Among the main air pollutants, carbon dioxide (CO_2), ozone (O_3), sulphur dioxide (SO_2) and fluorides (F-) can be counted.

Impact of Pollutants on Olive

1. Exposure to elevated CO_2 enhanced photosynthesis rates and decreased stomatal conductance, leading to greater water use efficiency.
2. Elevated O_3 doses (150 ppb) for a short exposure time (a few hours) led to a significant reduction in stomatal conductivity.
3. The SO_2 plays prominent role with other pollutants. Through the stomata, SO_2 can reach palisade and mesophyll cells, altering the functionality of stomata, photosynthesis, and leaf structure.
4. Long-term SO_2 treatments of olive plants determined genotype-dependent responses: on cv. Frantoio, they induced a strong reduction of the photosynthesis rate and stomatal conductance and, on several cultivars, the thickness of the leaf blade and the stomatal density and opening were reduced.

References

Azzi, G. 1928. Ecologia Agraria. U.T.E.T., Torino, Italy.

Bennet, W.F. 1993. Nutrient Deficiency and Toxicities. APS Press, St. Paul, MN, USA.

Ciccarone, A. 1956. Le malattie nutrizionali dell'olivo e relative rimedi. *Notiziario sulle Malattie delle Piante* **37-38** (N.S. 16-17): 71-89.

Connor, D.J., Fereres E., 2005. The physiology of adaptation and yield expression in olive. *Horticultural Review* **31**: 155- 229.

Fernández-Escobar, R. 1998. Diagnóstico del estado nutritive y fertilización del olivar. *Phytoma España* **102**: 54-55.

Gibelin, A.L., Déque M. 2003. Anthropogenic climate change over the Mediterranean region simulated by a global variable resolution model. *Journal of Climate Dynamics* **20**: 327-339.

Graniti, A. 1993. Olive scab: a review. *Bulletin OEPP/EPPO Bulletin* **23**: 377-384.

Gucci R., Mancuso, S., Sebastiani, L. 2003. Resistenza agli stress ambientali. In: Fiorino P. (ed.). Olea. Trattato di Olivicoltura, pp. 91-111. Edagricole, Bologna, Italy.

Mancuso, S. 2000. Electrical resistance changes during exposure to low temperature measure chilling and freezing tolerance in olive tree (*Olea europaea* L.) plants. *Plant Cell Environment* **23**: 221-229.

Messeri, A. 1947. Osservazioni morfologiche sulle "olive passerine". *Nuovo Giornale Botanico Italiano* (N.S.) **54**: 374-376.

Michailides, T.J., Vossen, P.M., McKenry, M.V. 2011. Diseases of olive in California. In: Schena L., Agosteo G.E., Cacciola S.O. (eds). Olive Diseases and Disorders, pp. 379-401.

Reuter, D.J., Robinson, J.B., Dutkiewicz, C. 1997. Plant Analy Analysis: An Interpretation Manual. CSIRO Publishing, Melbourne, Australia.

Valderrama, R., Corpas, F.J., Carreras A., Fernández-Ocaña A., Chaki M. Luque F., Gómez-Rodríguez M.V., Colmenero- Varea P., del Rio L.A., Barroso J.B. 2007. Nitrosative stress in plants. *FEBS Letters* **581**: 453-461.

28

Major Production Problems in Olive

The olive trees are xerophitic in nature and live more than 100 years of age. In Indian sub-continent a wild olive, *Olea cuspidata* is found wild in the North-West Himalayas and other adjoining hills do not produce quality fruits and are very poor in oil recovery. Cultivated olive *Olea europea* is commercially grown in areas have Mediterranean climate like Spain, Italy, Egypt, Algeria, France, Greece, Turkey, Cyprus, Israel, Jordan, Morocco, Palestine, Syria, Saudi Arabia, Tunisia, Lebanon, California and Bermuda. Several varieties were introduced in India from Italy, Spain, Egypt and Israel, planted at suitable places available in parts of Jammu and Kashmir, Himachal Pradesh, Uttarakhand and Rajasthan. The olive cultivation in India, witnessed several ups and downs at almost all the locations where the crop was introduced. However, initial results were found promising from all the three temperate states, J&K, HP and Uttarakhand. But in the later years of plant growth, the three remained unproductive and found unprofitable to growers. Therefore, it was not expanded in other parts of these states due to uneconomical venture. However, it is being still grown on smaller scale for oil extraction purpose at various pockets of these states. The low productivity of fruits and poor oil recovery is attributed by several factors like alternate bearing, pollination problems, biotic and abiotic stress, poor water and nutrient management *etc.* However, alternate bearing considered one of the major problems related to low productivity in olive.

Characteristics of Olive Cultivation

- ☆ Olive is a perennial fruit crop and a plantation takes between five to seven years to become fully productive.
- ☆ Production varies greatly and depends on the biological alternation of the olive tree (alternate bearing), farming methods (irrigated or rainfed), olive varieties (drought tolerant and drought sensitive) and the soil and climate conditions.

- ✰ Plantings of olive trees in marginal areas with poor productivity (mountainous or hilly areas); they can grow in poor, stony soil which it would be difficult to put to other crop uses.
- ✰ Olive trees are playing an important environmental role (fixing soils, biodiversity, and landscape).

Sustainable production in olives is the prerequisite for attracting the growers. The thorough understanding of the problems helps to suggest corrective measures and convert unprofitable to profitable venture. The salient features of these problems discussing as below:

1. Alternate Bearing

Alternate or biennial bearing severely affects olive production. Under these circumstances trees give heavy fruit crop one year followed by a light crop the next. This tree behavior is mainly due to flowering load and it is a feature to a greater or lesser extent of all cultivated olive varieties, and can be limited to an individual tree in an orchard or individual parts of a single tree. Goldschmidt (2005) suggested three principal regulatory mechanisms involved in the induction of alternate bearing: i) hormonal control, ii) nutritional (carbon and mineral balance) control and iii) flowering-site limitation. The synergistic and complementary effects of these three regulatory mechanisms are commonly used to reduce alternate bearing *i.e.*, through limiting fruit number (fruit thinning) provides additional flowering sites (Dag *et al.*, 2010), minimizes the amount of fruit-released inhibitors of early reproductive processes (Lavee, 2007) and prevents extreme nutrient depletion (Goldschmidt, 2005). However, the extent to which each mechanism affects alternate bearing differs largely among species and cultivars of fruit trees. Furthermore, the cause-and-effect relationships among these mechanisms with respect to alternate bearing are often unclear.

Causes of Alternate Bearing

There are two types of the situations may cause alternation: i) an off-year caused by a lack of flowers, a poor fruit set, or excessive drop and ii) an on-year with excessive fruit set, too little fruit drop, and too large a crop. These situations are the outcome of the several processes related to biotic and abiotic.

A. Environmental Factors

i) Climatic Stresses

Frost and low temperature: Spring frost, which destroys bloom, leads to alternation. Frost, an environmental factor, is unmistakably the trigger that starts a self-perpetuating process, driven by interacting plant activities. The process can be suppressed and tree behaviour corrected by chemical thinning. Synchronization of alternation over wide areas often has been blamed on spring frost with apples, olives (Morettini, 1950), pecans (Sparks, 1975), and mangos in cool areas (Singh *et al.*, 1974). Low temperature can also influence fruit set very strongly. Similar difficulties with higher temperatures are probably responsible for small crops of

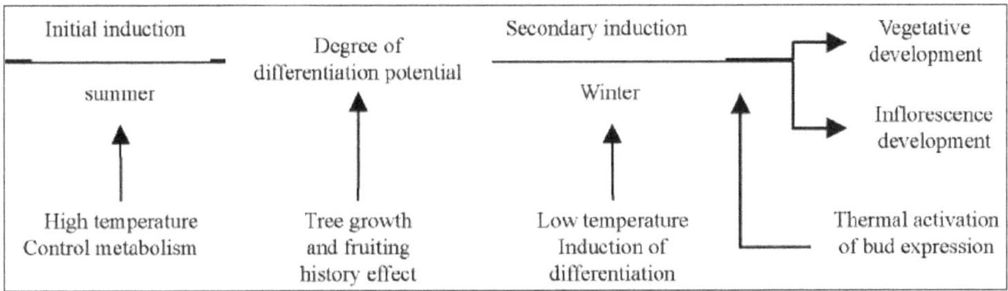

Figure 13: A Flow Diagram of the Effect of Environmental Conditions on the Stages Leading to Reproductive and Vegetative Development in Olive Trees (Lavee 2007).

olive, avocado and mango in other areas. Poor set may be due to low temperature effects on set itself or on bee activity.

Humidity: The environmental humidity affects yield through an excessive fruits drop (Chandler, 1950; Morettini, 1950). Low humidity cause low pollen germination due to desiccation of the stigma and may also be connected with cool night temperatures in arid zones. Low air humidity and other environmental stresses may also have indirect effects on set by enhancing leaf senescence and causing premature drop.

ii) Edaphic Factorts

Saline conditions cause leaf drop and are a possible indirect cause to the reduction of available reserves. During flower formation in olives, soil moisture stress is conducive to leaf abscission and to a high percentage of sterile flowers. Stressed olive trees produce about 47 per cent flowers per inflorescence, 22 per cent perfect flowers, and 9 per cent fruits as compared with unstressed (Hartmann and Panetsos, 1961). Summer drought also reduces fruit and oil yields. Drought also enhance drop of reproductive organs and leaves and reduce vegetative growth. However, irrigated olive trees yield better and regularly.

iii) Biotic Factors

Pests and diseases, attacking flowers, young fruits, leaves, and woody structures, can have a direct or indirect effect on alternation.

B. Endogenous Factors

The endogenous factors include the physiological and biochemical reactions occur as well affects the flower induction and related factors.

i) Inhibitory Effect of Frowing Fruits to Flower Initiation

Strong inhibition of flower bud initiation is occurs in deciduous trees where flower initiation occurs during the first stages of fruit development. The physiological processes underlying these effects have not been fully identified. After successful pollination fruits usually contain many developing seeds. Seeds usually increase the number of persisting fruits and reduce self-thinning. Both effects are due to enhanced production of growth regulators and to strong sink activity. Auxin produced by seeds and its movement is stronger for a biennial cultivar than for a regular bearer. It is quite possible that different hormonal and nutritional factors combine to depress flower formation when a considerable crop of seeded fruits develops, thus starting an endogenous alternating cycle.

ii) Pollination Related Problems

Most olive cultivars are self-incompatible or partially self-compatible and need to be fertilized by compatible pollinizer varieties to ensure a commercial yield. As per the reports Koroneiki is a self-compatible cultivar. However, varieties like 'Arbequina', 'Kalamata', 'Leccino', 'Manzanillo', 'Picholine' and 'Picual' are self-incompatible or partially self-compatible cultivars. 'Ascolana' and 'Mission' are cross-incompatible with 'Manzanillo' and 'Barouni' with 'Sevillano'. Cross-incompatibility has been reported to be reciprocal or bidirectional in some pairs of cultivars such as 'Manzanillo' and 'Mission'. Therefore, lack of suitable pollinizer, insufficient overlapping of blossom periods, and low activity or lack of interest on the part of pollinating insects may all cause poor yields.

iii) Vegetative to Reproductive Growth Balance

In olive, leaves plays important role in two ways, provide nutritional and regulates hormonal balance in reproductive organ development. However, it is difficult to discriminate between these two factors because both classes of compounds are often produced in close proximity and translocated (cytokinins and water-soluble gibberellins excepted) through the same vascular system. The contribution of leaves to flower bud formation has been established in most plants, including tree crop such as olives (Hackett and Hartmann, 19641). Other effects of leaves at later stages of flower formation include the special case of leaves antagonizing abscission of flower buds caused by:

The effect of fruit on flower bud induction: In addition to the effect of fruit on the level of vegetative growth, the developing fruits were shown to have also a significant effect on the development of flower buds for the following season. In olive this relation is not clear and seems to be not particularly significant.

Competition between vegetative and reproductive sinks: Seeds in growing fruitlets are usually considered a powerful sink favouring better mobilization of photosynthetic products by the growing fruit. It has also been emphasized that a delicate balance between fully vegetative and reproductive branches is needed for the regular cropping of olives (Poli, 1979).

vi) Role of Fruit Overload on Bearing

The heavy crop produced during the on-year is perhaps the most universally recognized cause of alternation. While several specific interactions between fruits and other organs have already been discussed, the general effect of fruit overload on tree physiology deserves further attention. The population of developing fruits creates a cumulative sink which requires a continuous supply of building materials. There is a large number of small sinks at fruit set; at the later stages

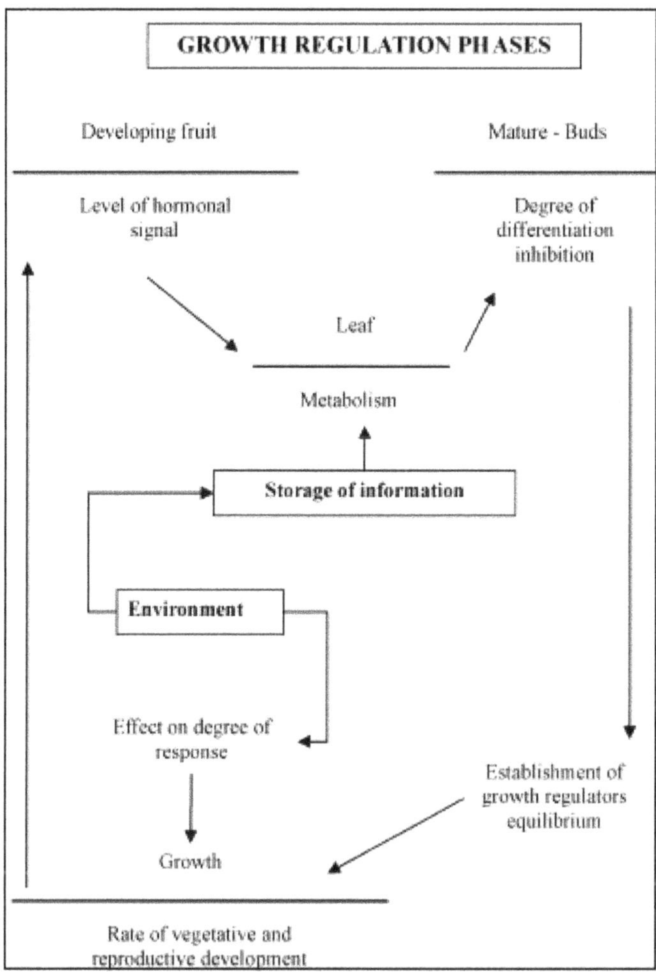

Figure 14: A Flow Diagram of the Stages Involving Growth Regulators in Controlling the Level of Alternate Bearing in Olive Trees (Lavee, 2007).

of fruit development there is a smaller, finite number of progressively larger organs individually demanding heavy investment of materials. Delayed harvest increases the interference with future flower bud production. Both mineral and organic nutrients may be obtained from either newly assimilated materials or reserves previously accumulated in different tree tissues. Depletion of reserves as a consequence of the on-year has been demonstrated in numerous instances and causes true collapse in extreme cases.

Management Practices to Reduce Alternate Bearing

Due to the high dependence of the process leading to flower bud differentiation and fruit set on the environmental conditions, there are regions where alternate bearing can only be partially overcome. Various methods have been developed and horticultural techniques modified to reduce or overcome alternate bearing:

i) Pruning

Pruning is one of the oldest methods to control production in olive orchards. It enhances light penetration through opening the trees for effective light penetration in to canopy to facilitate and assist for enhancing flower bud differentiation. It also helps in reducing the number of fruits by limiting the amount of fruiting wood. Pruning induces new vegetative growth at the stamps of the pruned branches and balances the shoot to leaf ratio. Application of sufficient resources (water and fertilizer) during the "on" years so that they can produce good quality shoot growth despite the heavy crop, and reduce the inputs during an "off" year.

ii) Fruit Thinning

Fruit thinning plays an important role on both fruit quality during the 'on' year itself and the fruiting potential for the following one, by reducing fruit number on the trees reduces the number of seeds and minimizes their inhibiting effect on the fruiting in the following season. The severe thinning is a useful tool to reduce alternate bearing particularly in regions with unstable production. Thinning is performed by spray of NAA usually 10-20 days after full bloom once the degree of fruit set is useful in olive. Late summer application of gibberellic acid shown to reduce flower bud differentiation and could be used to reduce the flowering towards the 'on' years.

iii) Harvesting Time

The slight effects on alternate bearing are reported when fruit harvest at early stages of maturation. However, there are negative effects if fruits harvested at advanced or full fruit maturation leading to alternate bearing. Therefore, avoid late harvest in the on-year, is critical both for oil and table olives, shifting in on years the product of the later toward early harvested fruit for green pickles.

iv) Girdling

Girdling is an efficient and feasible method for the reduction of alternate bearing in intensively cultivated olive orchards. Girdling increased fruit set and in some regions, mainly with warm winters, increased also the number of inflorescences when executed prior to the off year. It helps to increase the number of perfect flower on the girdled scaffolds, widening the ration between perfect and male flowers. Girdling might cause scaffold decline when applied to weak and slow growing trees under extensive cultivation without irrigation in stress prone regions.

v) Fruit Thinning

Olives are primarily valued based on fruit size. Heavy crops have preponderance of small fruit that are expensive to pick and of little value. Further, olives are alternate bearing; light crops invariably follow heavy crops. Therefore thinning is practiced to:

- ☆ ***Increase fruit size***: Removing a portion of the crop results in increased size for the remaining crop. The earlier the crop is thinned, the greater the

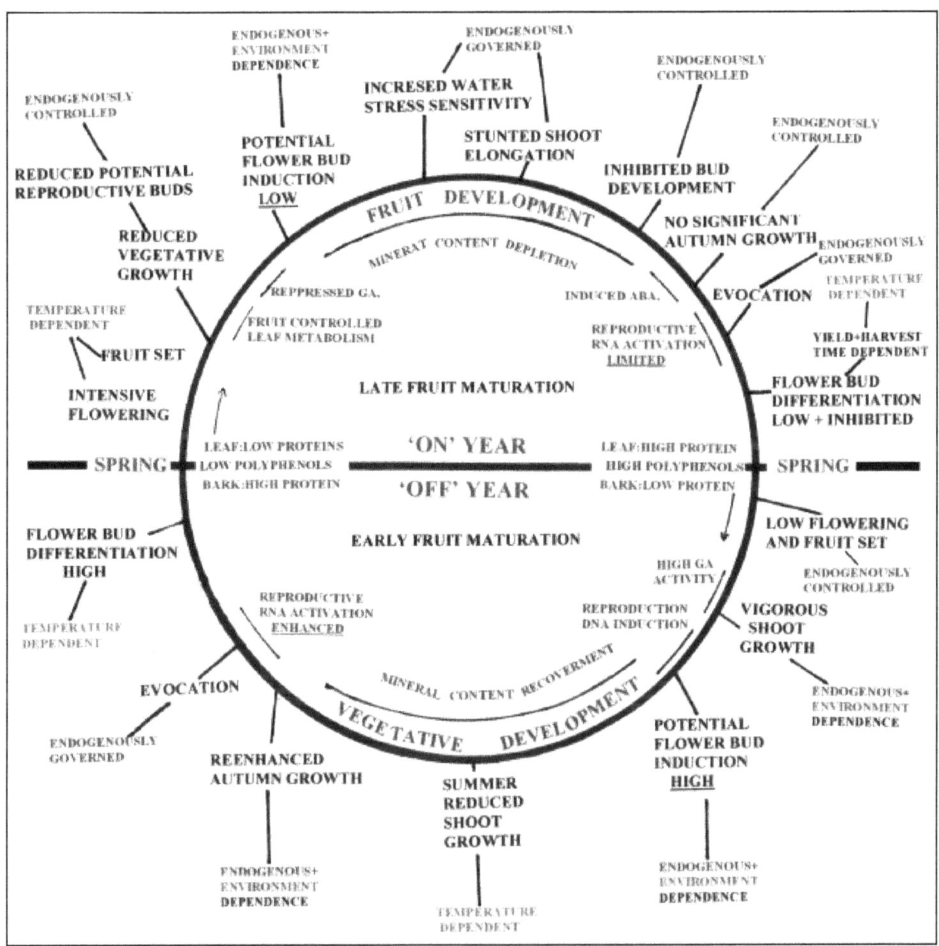

Figure 15: A General Scheme of the Alternate Bearing Events, Involvement of Endogenous Processes and Interaction with the Environment during an 'On' Followed by an 'Off' Year (in black are the developmental stages, in red endogenous processes and in green environmental involvement) (Lavee, 2007).

increase in fruit size will be. Since fruit buds begin forming during the growing season of the year previous to the appearance of the fruit, some thinning is done when the trees are pruned. Frequently, more thinning needs to be done once fruit development begins.

- ✰ ***Obtain continuous annual production*:** Alternate bearing is a sequence where a heavy crop develops one year, followed by a very light crop the next year. The degree of alternate bearing will vary among fruit crops and among varieties of the same type of fruit.
- ✰ ***Improve fruit quality*:** Thinning should increase fruit quality both in terms of appearance and taste. Fruit colour will be better and taste will be improved when a heavy crop of fruit is thinned.

- ☆ ***Avoid tree breakage***: The excess weight of a very heavy crop can cause limb breakage or splitting, especially when limbs have weak crotch angles or are excessively long.

Stages of Thinning

Stage of thinning corresponding to the stage of fruit growth behaves differently. The brief details are given as under towards regularity. At:

- ☆ ***Heavy fruit set stage***: Thinning is particularly important when fruit is set in clusters. Fruit should be thinned even if only part of a tree has a heavy crop.
- ☆ ***Very young trees stage***: Fruiting can stunt the growth of young trees, thus preventing them from attaining enough size to ever have a good crop.
- ☆ ***Older trees***: As trees age, fruit size tends to decrease. Thinning the crop, even if fruit set is only moderate, may be essential to attain good fruit size.
- ☆ ***Trees low in vigour***: Weak trees need to be thinned more severely than strong trees to attain good fruit size and to allow the tree to become stronger.
- ☆ ***Injured trees***: Thinning the crop allows the tree to devote more energy to repairing the damage.
- ☆ ***Trees on poor soil***: Trees on poor soils cannot support a heavy crop as well as trees on better soils. Thinning the crop heavily will encourage good fruit growth on the remaining fruit.
- ☆ ***Varietal variation***: Certain varieties will not develop good-sized fruit unless the crop is thinned early and heavy.

Method of Fruit Thinning

Depending on the type and the variety of fruit, fruit clusters should be broken up, leaving a single fruit. Fruit should be spaced from 6 to 8 inches apart on limbs using one of the following methods:

- ☆ ***Hand thinning***: Hand thinning is the most precise way of adjusting a crop load on a tree. Fruit clusters can be broken up, leaving just a single fruit, and fruit can be evenly spaced about 6 to 8 inches apart on limbs. Damaged or weak fruit can be removed during the thinning operation, leaving a better quality crop on the tree. Hand thinning is often delayed until the largest fruit is about the size of a nickel and the natural drop period for fruit has passed. Unfortunately, hand thinning is often done too late to achieve the maximum benefit and is not practical for olives.
- ☆ ***Jarring limbs or clusters***: "Bumping" or "jarring" limbs will dislodge some of the fruit and accomplish most of the needed thinning. A padded pole or a plastic bat works well. With them, limbs can be struck hard enough to knock off some fruit, yet not so hard that branches will be damaged. Thinning in this manner is much quicker than hand thinning, though not

as precise. It may be necessary to follow up thinning in this manner either by hand thinning to break up clusters of fruit or by using a stick with a short piece of hose on the end to strike the cluster and dislodge some of the fruit. Best results from thinning using this method will be obtained in early morning or late evening when fruit stems are firm. Thinning in this manner is generally done when the largest fruit are between the size of a dime and a nickel.

- ☆ ***Chemical:*** Chemical thinning is not an option for most non-commercial growers. It is a very precise operation, and either no thinning or overthinning can result. Good results depend on selection of the correct chemical, rate and time of application. Weather conditions before, during and after application can influence the results. When done properly, chemical thinning gives the best results of all the methods used, since it is done earlier than the others. Napthalene acetic acid (NAA) is an effective fruit thinner when applied 7 – 14 days following full bloom. Activity of the thinner is influenced by tree condition, air temperature, and NAA dose.

References

Chandler, W.H. 1950. Evergreen orchards. Henry Kimpton, London.

Dag, A., Bustan, A., Avni, A., Tzipori, I., Lavee, S., Riov, J. 2010. Timing of fruit removal affects concurrent vegetative growth and subsequent return bloom and yield in olive (*Olea europaea* L.). *Scientia Horticulturae* **123**: 469–472.

Goldschmidt, E.E. 2005. Regulatory aspect of alternate bearing in fruit trees. *Italus Hortus* **12**: 11-17

Hackett, W.P. and Harmann, H.T. 1967. The influence of temperature on floral initiation in olive. *Physiol. Plant* **20**: 430-436.

Lavee,S. 2007. Biennial bearing in olive. *Annals* **17**(1): 101-112.

Morettini, A. 1950. Olivicoltura. 1st Ed. Ramo Editoriale degli Agricoltori, Rome

Poll, M. 1979. Etude bibliographique de la physiologie de l'alternance de production chez l'olivier (*Olea europaea* L.). Fruits **34**: 687-694.

Singh, R.N., P.K. Majumder, P.K. Sharma, G.C. Sinha, and P.C. Bose. 1974. Effect of deblossoming on the productivity of mango. *Scientia Hart.* **2**: 399-403.

Sparks, D. 1975. The alternate fruit bearing problem in pecans. *Proc. Northern Nut Grow. Assoc.* **65**: 145-157.

29

Status of Olive Growing in India

In India,olive is grown only in small area (about 400 ha),mainly in the Himalayan mountainous region encompassing Jammu and Kashmir, Himachal Pradesh and Uttarakhand hills at an altitude ranging from 1000 to 1300 m above mean sea level. The Jammu and Kashmir state occupies maximum area (276 ha) which is very sporadic found in districts of Doda, Udhampur, Rajouri, Poonch, Kupwara, and Baramulla. The district Doda has the maximum area (59.76 per cent) followed by Udhampur (15.5 per cent). Wild Olive is reported from the Garhwal region of the North-West Himalaya. Indian wild Olive cultivar Olea ferruginea is also reported from Almora hills. The tree is somewhat localized in distribution and occurs chiefly along the outer hills and inner dry valleys of Western Himalaya. In India, the demand of olive is increasing very fast due to its peculiar medicinal properties. Most of the oil is still used for massage purpose but interest is now increasing to use it as edible oil. India imported about 1500 tonnes of olive oil in 2006 and 2300 tonnes in 2007 with growth rate of 53 per cent annually, which is expected to reach 42218 tonnes by 2012.Besides consumption for domestic market, olive in India has a tremendous potential for the following reasons because olive can be grown in less fertile soil than annual oil crops with higher productivity and little care in hilly regions which has ideal climate and contributes nearly 1/8th of total area of the country ranging from NWH to NEH region. Secondly, its seeds and pulp have potential to yield oil ranging from 40-70 per cent depending upon variety or hybrid in comparison to annual oil crops with a productivity of much higher (2.0 t/ha) as compare to just 0.5 t/ha in annual oil crops and thirdly oil priced between Rs. 400-500 per litre as compared to Rs. 80 per litre of annual oil crops. Olive thus has a better potential both in terms of productivity (4 times) and price (5 times) and can fit very well under marginal areas of mid hill regions of India. Its cultivation therefore can help to elevate poverty of small farmers of hilly regions. Besides oil some varieties can be grown for table and pickle purposes which have huge demand in the domestic and international market. Their production both for oil and table purposes by increasing the area under olive in mid warm temperate regions not only augment to our oil requirement but can also save foreign exchange to the tune of about 200 crores by reducing the import of edible oil from other nations.

Causes of Low Productivity

1. Poor orchard management with no or very less pruning and training, nutrition, irrigation, plant protection *etc.* resulting in very poor yield.
2. Excessive hot winds during flowering dry the stigma resulting in low fruit set and yield.
3. High annual rainfall during monsoon coincides with early fruit set, resulting in fruit drop.
4. Non-availability of water during winters and early spring.
5. Inadequate number of region specific high yielding varieties with suitable pollinizers and cross compatible combinations.
6. Low temperature during flower bud development resulting in ovule abortion.
7. The phenomenon of alternate bearing causing inconsistent productivity.
8. Lack of standard region specific production and protection technologies.
9. Inadequate processing and value addition facilities in the production sites.
10. Longer pre-bearing phase and lack of standard root stocks for specific purposes like high density planting, moisture stress *etc.*
11. Non availability of adequate quantity of quality planting material.
12. Limited support services for large scale commercial production.

Olive Promotion Projects

1. **Indo-Italian project:** Considering the importance of the crop an Indo-Italian project for promotion of olive was implemented in 1984 at two phases from 1984-87 and 1990-93 at all the three Northern hilly states of India *viz.*, J&K, H.P and U.P. But due to several constraints the real potential of the crop could not be harvested except few successful attempts around Ramban, Doda and Uri areas of J&K state.
2. **Horticulture Technology:** To improve the productivity, several experiments under HTM (MM-I) were carried out especially on rejuvenation and unfruitfulness. The results obtained indicated highly encouraging results. The light pruning, application of optimum dose of NPK fertilizers and 0.5 percent borax spray resulted in very good fruit set and changed unfruitfulness in existing varieties. Top working with side veneer grafting with improved cultivars on old varieties also gave very good results.
3. **Three-way collaborative project among Rajasthan–Israel and Indian firm:** Recently a pilot project on 250 hectares of land with 1.25 lakh olive

Table 25: Major Institutions Working on Olive

Sl.No.	Institutions	Nature of Work
1.	Central Institute of Temperate Horticulture	1. Exploration and germplasm conservation 2. Introduction of exotic and indigenous germplasm and their evaluation and identification. 3. Standardization of propagation techniques and root stocks and large scale multiplication. 4. Standardization of production and protection technologies on pollination, irrigation, nutrition, canopy management *etc.* 5. Processing and value addition.
2.	Dr. Y.S. Parmar University of Horticulture and Forestry, Solan, Himachal Pradesh	1. Introduction of exotic and indigenous germplasm and their evaluation and identification. 2. Standardization of propagation techniques and large scale multiplication. 3. Standardization of production and protection technologies. 4. Processing and value addition
3.	Sher-e-Kashmir University of Agricultural Sciences and Technology-Jammu, J&K	1. Standardization of propagation techniques and large scale multiplication. 2. Standardization of production and protection technologies.
4.	Sher-e-Kashmir University of Agricultural Sciences and Technology-Kashmir, Srinagar, J&K	1. Standardization of multiplication techniques through semi-hand wood utilizing. 2. Floral biology and nutritional studies in olive.
5.	Advance Centre for Horticulture,Ramban, Department of Horticulture, Government of J&K	1. Standardization of propagation techniques and large scale multiplication. 2. Processing and value addition
6.	Rajasthan Olive Cultivation Limited (ROCL), Durgapura, Jaipur, Rajasthan	1. Introduction of exotic germplasm from Israel. 2. Large scale multiplication of planting material. 3. Standardization of production technologies. 4. Processing and value addition.

trees of three varieties in dry desert Rajasthan have been planted under the technical expertise of Israeli firm by an Indian firm. As per the agreement, Indolive provided the expertise, Finolex the micro-irrigation solutions and equipment, and the Rajasthan government land and funds. RSAMB and Finolex invested Rs 1.50 crore each. About 112,000 olive plants of seven varieties were brought from Israel. Their rooted cuttings were hardened at a high-tech nursery set up especially for the project in Durgapura, near Jaipur. After the cuttings grew into small plants, they were transplanted in agriculture training centres (ATCs) at Sriganganagar, Nagaur, Bikaner, Jalore, Jhunjhunu, Alwar and Jaipur districts.The trees started bearing flowers and fruits in 2012, a year or two after they should have done so. Of the seven varieties planted on 182 hectare (ha), only three have yielded fruits, that too very little. The total production was only 10 tonnes and the oil yield was a mere 9-14 per cent of the olive production.

Strategies for Olive Promotion and Higher Production

1. Identification of suitable pockets for promotion of olive cultivation. Ideal temperature for growth and yield is 12-15°C from Oct-May. Areas fulfilling chilling and having optimum temperature range need to be identified for area expansion and promotion.
2. In NW Himalayan region there is a tremendous genetic diversity. A comprehensive exploration and improvement programme need to be under taken to identify the ideal indigenous cultivars.
3. Introduction, evaluation and identification of elite region specific genotypes.
4. Standardization of effective propagation techniques mass multiplication of elite genotypes and their large scale distribution among farmers.
5. Suitable training and pruning system fitting to our agro-climatic situation need to be standardized.
6. Since the crops is self-incompatible suitable pollinizers for commercial variety need to be standardized and planted depending upon the pollen production potential of pollinizer variety.
7. Horticulture techniques like top working and architectural engineering need to be undertaken on already existing olive plantations with improved varieties and for maintaining better canopy and light interception. This technique can also be employed for those orchards which are lacking in appropriate pollinizer cultivars.
8. The problem of prolonged juvenility need to be addressed in combination with suitable rootstocks, precocious varieties and effective growth regulators.
9. A complete site specific agronomic package need to be developed. Including crop protection.

10. Crop phonological studies required in relation to plant growth and development with focus on physiology and reproductive behaviour including floral biology.
11. There is need to standardize the nutritional and water requirements under different agro climatic zones suitable for olive cultivation.
12. Standardization of agro-techniques for the high density orcharding of olive with proper training, pruning, integrated nutrient management, drip irrigation and mulching for water conservation.
13. Post harvest handling, processing, packing, packaging, storage and value addition needs attention.
14. In order to enhance/ promote the cultivation of olive in the non -traditional areas a clear by back policies and processing centers need to be addressed.

30

Olive Cultivation in Rajasthan

In the desert state Rajasthan, cultivation of olive began in 2007 on experimental basis. In year 2015, Rajasthan is cultivating olives in state-farm fields in an area of about 240 hectares as part of the Indo-Israel project. In future, Rajasthan Government is likely to expand in 200 ha at farmer's fields through National Mission on Oilseeds and Oil Palm (NMOOP) and 5,000 ha under the Rashtriya Krishi Vikas Yojana (RKVY). The ROCL will provide the necessary saplings to selected farmers at 75 per cent of the plant cost.

Milestones of Olive Cultivation in Rajasthan

- ☆ In May 2006, Rajasthan chief minister Ms. Vasundhara Raje went to Israel as part of Indian Delegation to attend the 16th International Agriculture Exhibition and visited world famous Kibbutz, is a collective community in Israel that traditionally based on agriculture, in the Negev desert of southern Israel. She was impressed with the lush olive trees growing in the arid landscape.
- ☆ In June 2006, she took decision to replicate the programme in desert state.
- ☆ After six months, a tripartite agreement was signed between Israeli company Indolive Limited, Rajasthan State Agricultural Marketing Board (RSAMB) and Pune-based Plastro Plasson Industries India, now known as Finolex Plasson Industries (India) Ltd, to set up a joint venture and formed "Rajasthan Olive Cultivation Limited (ROCL)" in April 2007.
- ☆ The technical expertise on olive cultivation is provided by Indolive; micro-irrigation solutions and equipment made available by Finolex, and the Rajasthan government provided all the logistics like land and funds.
- ☆ In 2007, ROCL imported 112,000 olive saplings from Israel and planted on an area of 182 ha at seven locations *i.e.*, government farms situated at

Barore (Sriganganagar), Bakalia (Nagaur), Lunkaransar (Bikaner), Santhu (Jalore), Basbisna (Jhunjhunu), Tinkirudi (Alwar) and Bassi (Jaipur) of different agro-climatic regions of the state. The seven leading varieties were planted includes, 'Arbequina', 'Barnea', 'Coronaiki', 'Coratina', 'Frontoio,' 'Picholine', and 'Picual' for cultivation in Rajasthan.

- The first plants were planted in March 2008 at Bassi. Olive plants at all plantation sites showed good growth and fruiting occurred this year at three farms, Barore, Bakalia and Lunkaransar, while flowering was observed at Basbisna, Tinkirudi and Bassi. Well-known olive experts in the world have been regularly visiting, inspecting and monitoring the progress in the farms and providing technical know-how, while Israeli expert Gideon Peleg has been based in India as a full-time technical consultant.
- In 2012, these orchards started bearing flowers and fruits. Out of seven varieties planted on 182 hectare (ha), only three have yielded fruits. The total production was only 10 tonnes and the oil yield was a mere 9-14 per cent of the olive production.
- Rajasthan is become the first state to have an olive oil refinery established in October 2014 at a cost Rs. 3.75 crore at Lunkaransar in Bikaner.
- In 2014, the state produced 100 tonnes olive fruits and extracted 10,000 litres of oil, which was marketed under the brand name 'Raj Olive Oil'.
- One hectare of farmland under cultivation can produce 2.5 tonnes of oil and 15 tonnes of fruit with 12-16 per cent profit, bringing gains of up to Rs350, 000 per hectare.
- Commercial cultivation of olives started in Rajasthan after success of pilot project. Olive cultivation is expected to fetch about 5 times the profits that the farmers in Rajasthan currently fetch from wheat on a hectare land.

Index

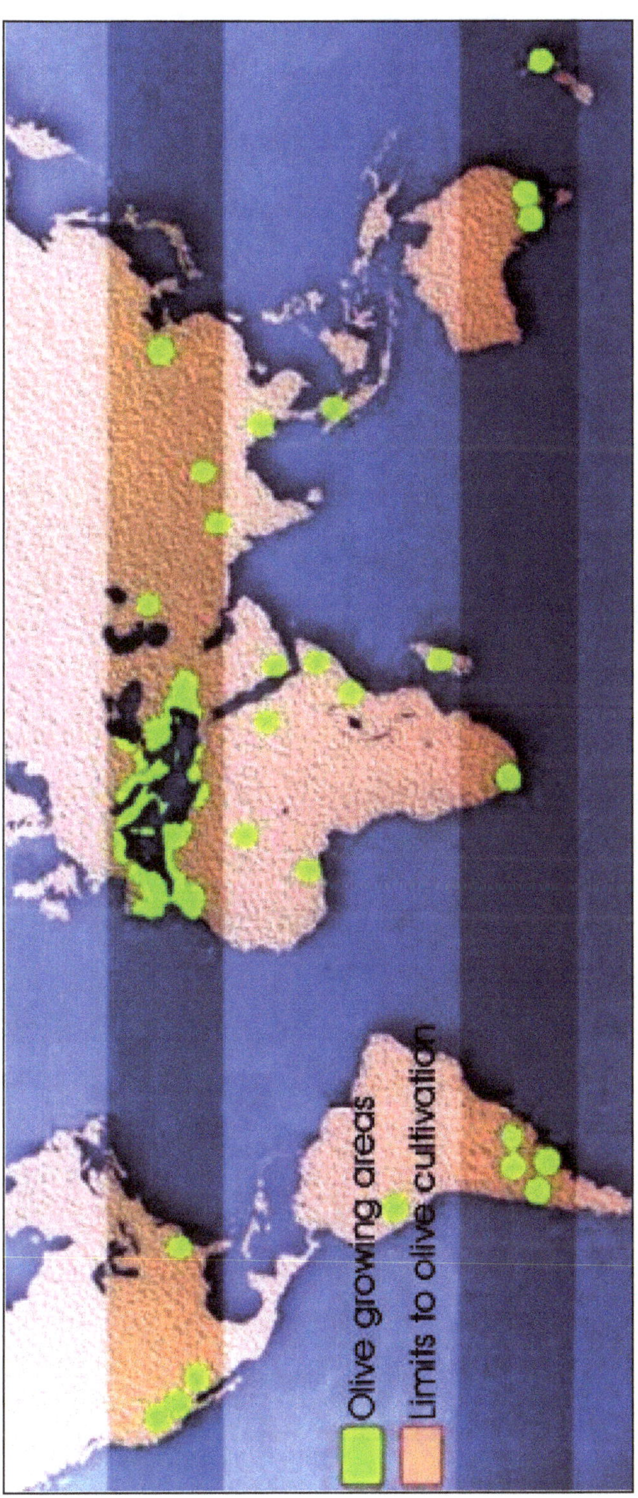

Figure 1: Olive Trees Distribution in the World.
***Source:* International Olive Council. (p. 5)**

(p. 63)

(p. 64)

(p. 65)

(p. 66)

(p. 67)

(p. 67) (p. 68)

(p. 69) (p. 70)

(p. 70) (p. 71)

(p. 71) (p. 72) (p. 73)

Figure 12: Optimum Fruit Harvesting Stage for Higher Oil Recovery in different Varieties of Olive. (p. 134)

www.ingramcontent.com/pod-product-compliance
Ingram Content Group UK Ltd.
Pitfield, Milton Keynes, MK11 3LW, UK
UKHW021010290726
14059UKWH00001BA/66

9 789386 071040